A Career in Air Traffic Control
(2nd Edition)

Michael S. Nolan
Sharon L. LaRue

A Career in Air Traffic Control, 2nd Edition
ISBN-13: 978-0-9962452-6-5
by Michael S. Nolan and Sharon L. LaRue

Distributed by:
eAcademicBooks LLC
115 E. 5th St.
Fowler IN, 47944
(312) 857-4071
support@eAcademicBooks.com

Author: Michael S. Nolan, Sharon L. LaRue
Layout: Laura A. Ballinger

About the Authors

Michael S. Nolan

Michael S. Nolan is the author of the textbooks Fundamentals of Air Traffic Control, Basic Aircraft Science and A Career in Air Traffic Control. Prof. Nolan currently holds the rank of professor and has taught at Purdue University since 1978. He holds a commercial pilot certificate as well as that of a certified flight instructor, airframe and powerplant mechanic, control tower operator and the national weather service limited aviation weather observer certificate.

Sharon L. LaRue

Sharon LaRue has taught air traffic control classes at the University of Alaska Anchorage for the past 15 years. She was previously an en route controller for ten years at Anchorage center, working both domestic and oceanic control. She also worked for the FAA as a training specialist and facility scheduler. Prof. LaRue holds a B.A. in English from the College of St. Benedict, and a M.A.Ed in Adult Education and Distance Learning from the University of Phoenix.

Dedication

I would like to dedicate this text to my family; daughters Linda and Erin, son David, and son-in law Dan. But most of all to my wife….without whom none of this would ever have happened!

Thank you all!

Michael

Acknowledgements

Michael S. Nolan

This book has its beginnings in the late 1980's when I found that there was no textbook in existence for air traffic control. Since I had to teach a course in ATC, I naively decided that I should write a book as well. Little did I realize it would still be around 30 years later!

It was published by Cengage for its first 25 years, but they have decided to get out of the aviation publishing business and returned the book back to me. I am working with a small, local publishing house now, trying some new techniques and ways to publish. I hope it works out for you. It's a living experiment. It will be fun to see how it turns out!

A book isn't written by an author, it's an entire community that helps. Over the years I have had lots of help....from within the FAA, the air traffic controllers, as well as educators and pilots. There are way too many people to list, and after 30 years, many have left their profession but their contribution lives on. If you provided information, reviewed the text, provided feedback on past editions, or used it as a student. I am indebted to each and every one of you.

Thank You

Sharon L. LaRue

I would like to thank my family for their patience and encouragement while completing this project, my colleagues at the University of Alaska Anchorage who always help me to be a better teacher, my colleagues around the nation at other CTI institutions, who understand the unique challenges of teaching this particular skill set and can always make me laugh, Mike Nolan for the opportunity to collaborate on his grand enterprise, and finally my students, who always remind me that the best way to learn about anything, particularly yourself, is to teach someone else.

Table of Contents

About the Authors ... **3**

Dedication ... **4**

Acknowledgements ... **5**

Air Traffic Controller: Flight Plan to a Career ... **13**

What is Air Traffic Control? ... 13

Careers in ATC ... 14

DoD employment ... 14

Contractor Operated Facilities ... 14

FAA Employment ... 15

Control Towers ... 15

Terminal Radar Approach Control (TRACON) ... 15

Enroute controllers ... 15

Getting Hired by the FAA ... 16

FAA hiring requirements ... 17

Medical Examination ... 18

Security Investigation ... 21

FAA Academy Training ... 21

Field Training Program ... 21

Salaries ... 22

ATC Facility Classifications ... 23

Air Route Traffic Control Center ... 23

Terminal Radar Approach Control ... 23

Different control tower configurations ... 24

Tower without Radar ... 24

Tower with Radar ... 24

Combined Radar Approach Control and Tower with Radar ... 25

Combined Nonradar Approach Control and Tower without Radar ... 25

Combined Control Facility (CERAP) ... 32

Federal Contract Air Traffic Control Towers ... 32

Flight Service Stations ... 37

Navigation, Airports, and the National Airspace System ... **39**

Visual Navigation ... 39

Aeronautical Charts ... 40

Sectional Charts ... 40

Instrument Flying ... 40

Navigation Systems ... 41

VHF Omnidirectional Range (VOR) ... 41

VOR Categories ... 42

Aircraft Positioning Methods ... 43

DME Position Determination ... 43

Tactical Air Navigation (TACAN) ... 44

VORTAC ... 44

Area Navigation ... 44

Global Navigation Satellite System ... 45

Global Positioning System ... 45

GNSS Augmentation ... 46

Satellite Based Augmentation Systems ... 46

Ground-Based Augmentation System (GBAS) ... 47

Aircraft Based Augmentation System ... 47
Required Navigation Performance ... 47
Instrument Approach Procedures ... 49
Segments of an Instrument Approach Procedure ... 49
Initial Approach Segment ... 49
Intermediate Approach Segment ... 50
Final Approach Segment ... 50
Nonprecision Approach ... 50
Precision Approach ... 50
Approach Navigation Aid Classifications ... 51
Instrument Landing System ... 51
Localizer ... 51
Glide Slope ... 52
Marker Beacons ... 52
ILS Categories ... 52
GPS-Based Instrument Approaches ... 55
Waypoints/Computer Navigation Fix ... 55
Approach and Landing Procedures ... 56
Airport Markings and Signage ... 59
Runway Layout and Numbering ... 60
Pavement Markings ... 60
Taxiways ... 61
Runway and Approach Lighting ... 62
Runway Lighting ... 63
Runway Light Intensity ... 63
Embedded-in-Runway Lighting ... 64
Taxiway Lighting ... 64
Approach Lighting Systems ... 64
Vertical Guidance Systems ... 66
Precision Approach Path Indicator ... 68

Air Traffic Control System ... 69
Airspace ... 69
Airspace Classifications ... 70
Airspace Classes ... 70
IFR Flight in Controlled Airspace (Class A, B, C, D, and E) ... 71
Aircraft Separation in Controlled Airspace ... 71
Air Traffic Control Procedures ... 71
Sectors ... 72
Hand off Procedures ... 73
Controller Duties in an Air Route Traffic Control Center ... 73
Flight Data Controllers ... 73
Radar Controllers ... 73
Radar Associate/Nonradar Controller ... 74
Air Traffic Control Tower Responsibilities ... 74
Flight Data ... 74
Clearance Delivery ... 74
Ground Control ... 74
Local Control ... 75
Approach and Departure Control ... 76
Flight Operations ... 77
Aircraft Flight Plan ... 77

ATC Clearance ... 78
ATC Services in Different Airspace Classes ... 78
Class A ... 78
Class B ... 78
Class C ... 80
Class D ... 82
Class E ... 82
Class F and G Airspace ... 82
Airways ... 83
Aircraft Altitude and Flight Levels ... 84
Airway Structure Improvements ... 85
Navigation Reference System ... 85
Special Use Airspace ... 86
Prohibited Areas ... 86
Restricted Areas ... 87
Temporary Flight Restrictions ... 88
Special flight rules areas ... 88
DC Flight Restricted Zone and Special Flight Rules Area ... 88
Warning Area ... 88
Military Operations Area ... 89
Military Training Routes ... 89
Alert Areas ... 90
Controlled Firing Areas ... 90
National Security Areas ... 90
Communications Systems ... 90
Radio Phraseology ... 90
Radio Frequency Assignments ... 91
Standard Phraseology ... 91
Standard Pronunciation ... 92
Numbers ... 92
Altitudes ... 93
Flight Levels ... 93
Time ... 94
Altimeter Settings ... 94
Wind Direction and Velocity ... 95
Headings ... 95
Radio Frequencies ... 95
Aircraft Speeds ... 96
Air Traffic Control Facilities ... 96
Airways, Routes and Navigation Aid Descriptions ... 96
Aircraft Identification ... 97
Surveillance ... 98
Radar ... 98
Antenna ... 99
Receiver ... 100
Radar Display ... 100
Video Map ... 100
Types of Air Traffic Control Radar ... 101
Airport Surveillance Radar ... 101
Air Route Surveillance Radar ... 101
Airport Surface Detection Equipment ... 102

Air Traffic Control Radar Beacon System 103
Code Selection 103
Computerized Radar Systems 104
Automated Radar Terminal System 105
ARTS-II 105
Standard Terminal Automation Replacement System 105
Enroute Automation Modernization 106

Air Traffic Control Towers 107
Local Control 108
Local Controller Duties 108
Runway Separation 109
SRS aircraft categories 109
Departing Aircraft Separation 109
Line up and wait 111
Intersecting Runway Separation 112
Arriving Aircraft 112
Landing Clearance 114
Go Around 115
Arrival Runway Separation 115
Arrival-Departure Separation 116
Intersecting Runway Separation 117
Land and Hold Short Operations 118
Wake Turbulence 120
Wake Turbulence Weight Classes 122
Ground Control 122
Preventing Runway Incursions 123
Flight Data 123
Clearance Delivery 125
Datalink Services 125
Flight Progress Strips 126

Approach Control 129
Basic IFR Separation 130
Flight Progress Strips 131
Radar Separation Procedures 132
Radar Identification 132
Transfer of Radar Identification 133
Handoffs and Pointouts 133
Radar Separation 134
Vertical Separation 134
Longitudinal and Lateral Separation 135
Visual Separation 135
Radar-Assisted Navigation 136
Radar Arrivals and Approaches 137
Traffic Flow in TRACON Airspace 137
Approach Control Operations 139
Arrival Controller Responsibilities 140
Approach Clearance 143
Minimum Vectoring Altitude 144
Departure Control 144
Overflights and satellite airports 146

Radar Traffic Information 147
Conflict Alert 147
Minimum Safe Altitude Warning 148

Enroute Air Traffic Control 149
ARTCC Organization 150
Domestic Enroute Air Traffic Control 151
Traffic Flows within an ARTCC 153
Preferred Routes 154
Controller Duties in an Air Route Traffic Control Center 155
Radar Controller 156
Radar Associate/Non-radar Controller 156
Traffic Management Unit coordinators 156
Flight Data and Radar Data Processing 156
ERAM 157
Trajectory Modeling 157
ERAM Display 158
Enroute Separation 158
Separation Procedures 159
Vertical Separation Criteria 160
Lateral Separation Criteria 160
Longitudinal Separation Criteria 161
Handoffs, Spacing, and Sequencing 163
Miles in Trail Restrictions 164
Aircraft Metering 165
Delay Techniques 166
International Air Traffic Control 166
Canadian Air Traffic Control 168
International Airspace 168
Airport Identifiers 170
MNPS Airspace Operations 172
Pacific Ocean Air Traffic Control 174
Advanced Technologies and Oceanic Procedures 175
Trans-Polar Flights 176

Future of Air Traffic Control 179
Future Growth in Air Traffic 180
Next Generation Air Traffic Control 180
Performance Based Operations 181
Performance Based Communication 181
Performance Based Navigation 182
Performance Based Surveillance 184
Next Generation Air Traffic Control 185
Trajectory-Based Operations 185
Flexible Airspace Management 186
Collaborative Air Traffic Management 186
Future Employment in Air Traffic Control 188
ATC Privatization 189

Figures and Attribution 191

Index 195

Chapter 1

Air Traffic Controller: Flight Plan to a Career

Air traffic control is an exciting, interesting, exacting, and high paying career open to anyone with a willingness to study, learn, and work hard. It can be a difficult profession to enter, but the rewards are worth it! This book is an attempt to inform you about all the different careers available as well as act as a primer concerning the basic principles and practice of air traffic control. It can't make you a controller, but it can make you a better-informed applicant or student of the profession.

What is Air Traffic Control?

Air traffic control, usually abbreviated as ATC, exists not only to provide for the safety of aircraft as they travel from airport to airport, but also to effectively and efficiently manage all the resources needed to operate the complex system of navigation aids (navaids), communications and traffic management systems, as well as air traffic control facilities. This combination of facilities is commonly known in the U.S. as the national airspace system or NAS.

Careers in ATC

There are many different jobs and career paths for someone interested in an ATC career. People usually associate ATC with the most visible type of facility, the control tower. (Figure 1-1). However, there are many other interesting, challenging and high paying jobs available to those interested in working in this area. Most are with the Federal Aviation Administration (FAA), which is the agency that coordinates and operates the NAS. But there are other employers such as the Department of Defense (DoD), private contractors, as well as municipalities and airport operators as well.

Figure 1-1. Air traffic control tower at Ronald Reagan Washington National

Figure 1-2. Potomac consolidated TRACON

More than 14,000 federal air traffic controllers work in airport traffic control towers, terminal radar approach control facilities and air route traffic control centers across the U.S. for the FAA. (Figure 1-2). An additional 10,000 or so civilian contract and military controllers also provide air traffic services all told staffing over 500 air traffic control facilities in the United States. Taking into account future retirements, the FAA projects that about 1,000 new controllers per year will need to be hired to replace controllers who either retire or choose to leave employment.

DoD Employment

Individuals interested in working for the military in positions of air traffic control go through the normal enlistment process and are trained for the job by the respective military service at special schools. Some DoD facilities are operated by civilian controllers hired as federal employees, with most of them gaining their experience and training through the military via prior service, some are ex-FAA controllers. However, virtually no employment opportunities exist for anyone without previous certification or experience at DoD facilities.

Contractor Operated Facilities

A number of smaller ATC facilities (primarily low activity control towers) are operated on a contract basis with one of three private companies. Similar to the DoD situation described previously, most contract controllers are either ex-military or ex-FAA controllers. Private contractors seldom hire non-certificated employees.

FAA Employment

The majority of controllers employed in the U.S. work for the FAA which is the federal agency responsible for the safety of civilian aviation operations in the United States. There are primarily three different types of ATC facilities operated by the FAA. Air traffic control tower (ATCT) controllers work in the glassed-in towers you see at airports. Terminal radar approach (TRACON) controllers work in large, semi-darkened rooms, sometimes located within the control tower or in a separate building. Air route traffic control center (ARTCC) controllers work in one of a number of large facilities located across the country. ARTCCs separate air traffic at high altitude or at low altitude between low-activity airports. They are also responsible for arrivals and departures at many airports without control towers. (Figure 1-3).

Figure 1-3. Jacksonville, Florida ARTCC

Control Towers

Air traffic controllers working in the glass-enclosed control tower cabs manage traffic within a radius of a few miles of the airport using both visual and electronic means. They instruct pilots during taxiing, takeoff and landing, and they grant clearances for aircraft to fly. Tower controllers ensure minimum separation distances between landing and departing aircraft, transfer control to radar controllers when aircraft leave their airspace and receive control of aircraft for flights coming into their airspace. They are also responsible for all aircraft operating on the runway and taxiway surfaces.

There are a variety of controller operating positions in the tower cab, such as local control, ground control, flight data, cab coordinator, etc. Depending on the airport layout and/or size of the tower cabs (some airports have more than one tower), there can be more than one of the same types of position on duty at any one time.

Terminal Radar Approach Control (TRACON)

Air traffic controllers working in TRACONs typically manage traffic within a 40-mile radius of the primary airport using radar. They control the airspace extending from the surface up to between 10,000 and 15,000 feet above the airport. They use a radar system to separate departing and arriving flights, and they grant clearance for aircraft to fly through the TRACON's airspace. These controllers use radar to ensure that aircraft maintain minimum separation while within their airspace, transfer control of aircraft to tower or enroute center controllers when the aircraft leave their airspace and receive control of aircraft for flights coming into their airspace.

TRACON airspace is usually divided into discrete sectors that provide services to either a single airport or to multiple airports. Consolidated or large TRACONs in major metropolitan areas provide service to several primary airports. Their airspace is divided into areas of specialization, each of which contain groups of sectors. Controllers can be assigned to various positions in the TRACON such as such as final (sequencing), handoff and/or departure control.

Enroute controllers

Air traffic controllers assigned to enroute centers (ARTCC) guide airplanes flying outside of or above TRACON airspace. They also provide approach control services to small airports around the country where no TRACON service is provided. As aircraft fly across the country, pilots talk to controllers at successive enroute centers.

Each center's airspace is divided into smaller, more manageable blocks of airspace called areas and sectors. Much like the tracon, enroute controllers are assigned to different operating positions within each group of sectors they are assigned. These positions include radar, radar associate, coordinator, and assistant controller. ARTCC's typically have responsibility for more than 100,000 square miles of airspace extending over a number of states.

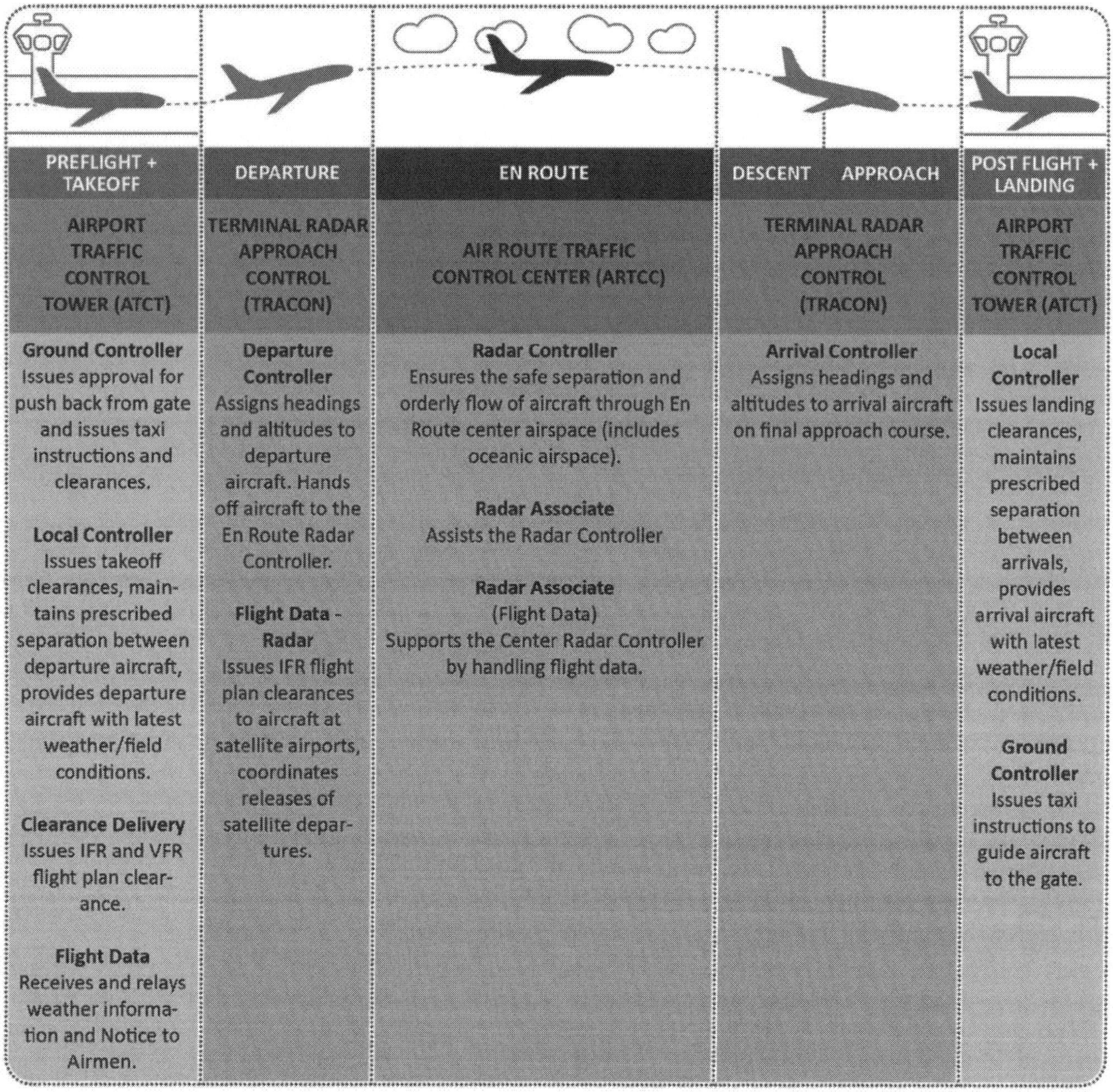

Figure 1-4. Facility operations overview by phase of flight

Getting Hired by the FAA

A potential controller must complete a number of steps and pass several tests before he or she can be considered for employment by the FAA. Many applicants begin the process of becoming a controller, but relatively few pass everything required to become fully certified and employed. In recent years, the FAA has modified its hiring and training process and plans to continue to make changes. It would therefore be in your best interest to visit the FAA website (www.faa.gov) routinely to keep abreast of any changes to the hiring process. The hiring processes and qualifications following describe the currently-existing process; it is subject to change with little or no notice from the FAA.

FAA Hiring Requirements

The minimum requirements for FAA employment as an air traffic controller include the following.

Applicants must complete all of the following steps:

- ✓ Be a United States citizen (naturalized is acceptable).
- ✓ Complete the application and hiring process and be able to start employment no later than their 31st birthday. (The FAA cannot and will not "speed up" your application if you are approaching this date cut-off).
- ✓ Have either three years of progressively responsible work experience, or a Bachelor's degree, or a combination of post-secondary education and work experience that totals three years, or have graduated from the AT-CTI institution.
- ✓ Pass an FAA medical examination.
- ✓ Pass a federal security investigation.
- ✓ Pass FAA air traffic pre-employment tests, which include both a biographical assessment as well as an ATC skills aptitude test. The biographical component is not required for AT-CTI graduates.
- ✓ Speak English clearly enough to be understood over communications equipment.

The FAA does not take direct applications for ATC positions. Interested applicants must apply via the government hiring website at www.usajobs.gov. When accessing this website, search for open positions using the term "air traffic control", or the position number 2152. The official job title for most openings in the FAA is Air Traffic Control Specialist. Typically, there are always a couple of open announcements for air traffic controllers on the website, but many are restricted to either certified controllers with previous experience or those with a military background. Those without air traffic control experience can only apply for positions advertised as open to all. For the purpose of the initial application and qualification, a college degree in ATC or a related area is NOT considered air traffic control experience, although that experience/education will likely prove to be very valuable if you are hired through the open application process.

Open application vacancy announcements for beginning air traffic control specialist positions are announced periodically (usually once per year) and the open period for applications is fairly short, normally about two weeks. If you miss the application period or are not selected for further processing, you may always re-apply during the next open period. Many successful controllers had to apply a number of times before finally being selected by the FAA for employment.

Figure 1-5. FAA hiring requirements for air traffic controllers

Medical Examination

As part of the hiring process, applicants selected for employment consideration are scheduled for an FAA provided medical exam. This examination includes a physical and eye exam, blood chemistry tests, an audiogram, a psychological test, and a drug screening. The physical examination is similar to the second-class medical exam required of commercial pilots.

Controllers are required to maintain their physical health throughout their career, which will be verified by regular medical exams. Controllers are required to undergo a physical once every two years until age 40, and then once every year until they retire. Some of the medical requirements required for employment with the FAA include the following:

Vision—Must have distant and near vision of 20/20 or better in each eye separately, without correction, or have lenses that correct distant and near vision to 20/20, each eye separately. Applicants must also have normal color vision.

Hearing Standards—No hearing loss in either ear of more than 25 db at 500, 1,000 and 2,000 Hz and no more than a 20-db loss in the better ear by audiometer, using ANSI (1969) standards.

Cardiovascular Standards—No medical history of any form of heart disease. A history of high blood pressure requiring medication requires special review.

Neurological Standards—No medical history or clinical diagnosis of a convulsive disorder, or a disturbance of consciousness, without satisfactory medical explanation of the cause, and must not be under any treatment, including preventive, for any condition of the nervous system.

Psychiatric Standard—Any medical history or clinical diagnosis of a psychosis or other severe mental disorders is disqualifying.

Diabetes—A medical history or diagnosis of diabetes mellitus is not automatically disqualifying but will require special review by the FAA.

Substance Abuse/Dependency—A history of substance abuse/dependency, including alcohol, narcotic, non-narcotic drugs, and other substances, will be extensively investigated by the FAA.

Psychological Exam—Individuals must take and pass a psychological exam.

General Medical—All other medical conditions will be evaluated on an individual basis. All applicants' medical histories and current examinations will be carefully reviewed. This includes past medical records and, if applicable, a review of military medical records.

Every applicant for employment as an air traffic controller is required to provide a urine sample during the medical exam that is screened for illegal drugs. The presence of drugs disqualifies an applicant for employment with the FAA. Random drug testing will occur throughout employment with the FAA. Failing one of these random drug tests can result in termination of employment. The use of marijuana is still a disqualifier for FAA employment. Federal law still considers marijuana to be an illegal drug, regardless of recent state initiatives and/or changes to state law.

MEDICAL REQUIREMENTS

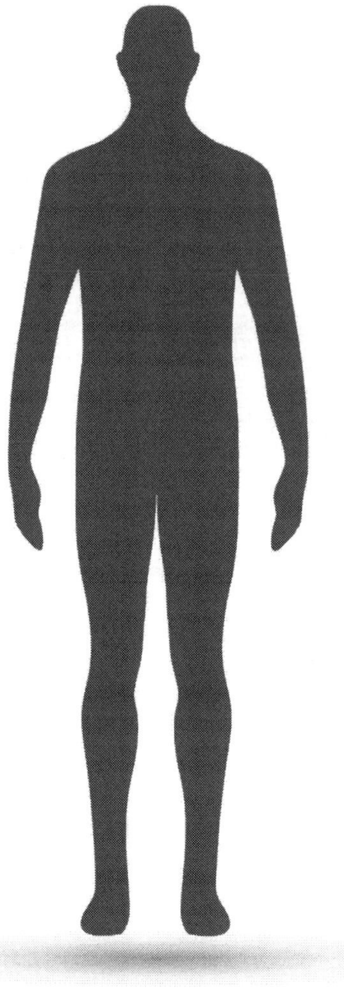

Hearing
No hearing loss in either ear of more than 25 db at 500, 1,000 and 2,000 Hz and no more than a 20-db loss in the better ear by audiometer, using ANSI (1969) standards.

Cardiovascular Standards
No medical history of any form of heart disease. A history of high blood pressure requiring medication requires special review.

Psychiatric Standard
Any medical history or clinical diagnosis of a psychosis or other severe mental disorders is disqualifying.

Psychological Exam
Individuals must take and pass a psychological exam.

General Medical
All other medical conditions will be evaluated on an individual basis. All applicants' medical histories and current examinations will be carefully reviewed. This includes past medical records and, if applicable, a review of military medical records.

Vision
Must have distant and near vision of 20/20 or better in each eye separately, without correction, or have lenses that correct distant and near vision to 20/20, each eye separately. Applicants must also have normal color vision.

Neurological Standards
No medical history or clinical diagnosis of a convulsive disorder, or a disturbance of consciousness, without satisfactory medical explanation of the cause, and must not be under any treatment, including preventive, for any condition of the nervous system.

Diabetes
A medical history or diagnosis of diabetes mellitus is not automatically disqualifying but will require special review by the FAA.

Substance Abuse/Dependency
A history of substance abuse/dependency, including alcohol, narcotic, non-narcotic drugs, and other substances, will be extensively investigated by the FAA. Marijuana is considered illigal.

Figure 1-6. FAA medical requirements for air traffic controller applicants

Security Investigation

Upon successful completion of the medical examination, the FAA will conduct a detailed security investigation of the applicant. This investigation includes an extensive civil and criminal background check, enquiries to former employers and educational institutions, and a review of any appropriate FBI, military, and police files. Some of the items included in the security investigation are military discharge history, any possible government loyalty issues, dishonesty in an application or examination process, any drug-related felony and/or firearms or explosives offenses, a history of alcohol-related incidents, willful disregard of financial obligations, derogatory employment terminations, or any other pattern or combination of incidents that lead to questions about applicant behavior and intent. Most security checks now include some form of scrutinizing social media as well, so investigating officials will be reviewing Facebook, Instagram, Snapchat, and other social media sites as well.

FAA Academy Training

Presently, anyone hired by the FAA as an air traffic controller receives their initial training at the FAA training academy located in Oklahoma City. Only individuals selected by the FAA for employment can attend this program. You cannot attend as a non-FAA employee. Those selected for training are considered FAA employees and are paid while attending, which is roughly three months for terminal hires, and five months for enroute hires. It is NOT possible to attend the FAA Academy unless you have been selected for employment. While training at the academy, you are paid as an employee, are eligible for all employee benefits, and are paid a stipend to help defray the cost of local housing during the training program.

The training at the academy focuses on the basic knowledge and skills required of an air traffic controller. Students learn about different aircraft types and operations and about the administrative and operational structure of the FAA. Air traffic control rules, regulations, and operational procedures are stressed. Practical experience through simulation is gained in control tower operation, nonradar, and radar separation. Students use computer-based training systems and computer-controlled control tower and radar simulators as an integral part of this unique training program.

The academy training program is generally composed of three parts: aviation academics, part-task training, and skills building. Performance is graded throughout the program, and at the conclusion of the academy training program, each student completes a series of performance verification exams, which determine a large portion of their final score. A final score with a percentage of 70 or higher is considered passing. Upon successful completion of these exams, the employee is converted to full-time status and placed at an appropriate facility for further training and certification. Anyone failing the academy training program is usually terminated from and no longer employed by the FAA. Presently, applicants are not afforded a chance to repeat the academy training program.

Field Training Program

After completing the academy training program, the newly-designated developmental controller is sent to an air traffic control facility for further training. Depending on the complexity of the facility, it may take between two and four additional years to become fully certified as an air traffic controller.

Developmental controllers typically begin training on an operating position such as flight data. After certifying on their first operating position, they begin to work as a controller, and simultaneously start training on the other positions at the facility. At a control tower, the training sequence is usually flight data, clearance delivery, ground control, local control, and the radar control positions. Center controllers begin at flight data and progress through radar associate/nonradar controller before certifying as a radar controller.

Throughout the training process, as developmental controllers are certified on more positions, they receive raises commiserate with their additional responsibilities. Once a developmental controller has certified on every position at the facility, they complete a facility rating exam. Upon passing this exam, they are considered facility-rated, certified professional controllers (CPC) and are paid accordingly.

Salaries

A controller's salary is based on the type and complexity of the facility to which he or she is assigned. Controllers at more complex facilities receive correspondingly higher salaries. ATC facilities are classified from a level 4 to a level 12. Upon completion of the FAA Academy training program, controllers are sent to their first field facility for on the job training. As controllers progress through training and meet facility-specific requirements, their pay grade is raised, step by step, from their developmental salary to the pay level assigned to that facility. Pay then increases from the minimum to the maximum base salary for each facility as a controller accrues years of service.

Once controllers reach CPC status at their facility, their pay moves up within the range specified for that facility. If a controller transfers to another facility, he or she typically retains the same salary, so long as it is still within the range of pay specified for the new facility.

Since many FAA facilities are open 24 hours per day, seven days per week, controllers usually work varying shifts. Their base pay is predicated on a normal 5-day work week. Controllers receive additional pay for working outside the "eight to five" hours that most people consider normal. All controllers earn Sunday premium pay at an additional rate of twenty-five percent (25%) of their hourly rate of pay. They also earn night differential at an additional rate of ten percent (10%) of their hourly rate of pay for all hours worked between 6 p.m. and 6 a.m. Overtime is paid at a rate of 150% (time and one half). Holiday pay is "double time" (200%). Controllers also receive premium pay of 10% for time spent as an on-the-job training instructor (OJTI) or when they perform controller-in-charge (CIC) duties. Controller's salaries are also affected by the geographic area in which they live. Those working in areas with higher costs of living may also be eligible for additional compensation called locality pay.

TABLE 1-1 ATC PAY LEVELS INCLUDING BASIC LOCALITY PAY (2018)

ATC Level	Minimum Base Pay	Maximum Base Pay
in training	$45,127	*
4	$58,683	**$79,221**
5	$66,604	**$89,915**
6	$73,600	**$99,358**
7	$81,328	**$109,792**
8	$89,867	**$121,323**
9	$99,301	**$134,056**
10	$114,198	**$154,167**
11	$120,189	**$162,258**
12	$126,202	**$170,374**

**While in training, controller pay climbs in proportion to the number of control positions in the facility certified to work. (25%, 50%, 75%). Once a controller is certified as a CPC, they earn the minimum base pay for that level of facility.*

Table 1.1 depicts the base annual salary for ATC controllers. This table includes a base level salary that applies to most locations. Controllers who work in areas with higher than average costs of living can expect to earn an additional 5%-15% to help defray the cost of living in one of these areas. As a general rule, based on the various salary differentials, most controllers can expect to make 10% to 35% more than the base salary shown in any given year.

As a controller works longer at any particular facility, their pay increases from the minimum towards the maximum for that facility. When transferring to another facility, pay and experience generally transfers with you. By the time you have worked roughly 20 years, you will likely be near the top of your facility's pay scale. Periodically, the entire pay scale is increased to take into account inflation and cost of living increases.

ATC Facility Classifications

The FAA operates over 300 air traffic control facilities in the United States. There are basically eight different types of types of ATC facilities; each assigned a pay level based on their traffic complexity and volume. It is possible for more than one type of ATC facility to be co-located in the same building, particularly if there is a tower and tracon facility.

Air Route Traffic Control Center

An ARTCC is an air traffic control facility that provides air traffic control service primarily to aircraft operating on IFR flight plans principally during the enroute phase of flight. ARTCC's always have a three-letter identifier that begins with a "Z". Following is a list of the existing ARTCC's:

TABLE 1-2 AIR ROUTE TRAFFIC CONTROL CENTERS

Identification	Location	State	Level
ZAN	Anchorage	AK	10
ZLA	Los Angeles	CA	12
ZOA	Oakland	CA	11
ZDV	Denver	CO	10
ZJX	Jacksonville	FL	11
ZMA	Miami	FL	11
ZTL	Atlanta	GA	12
ZHN	Honolulu	HI	11
ZAU	Chicago	IL	12
ZID	Indianapolis	IN	12
ZKC	Kansas City	KS	11
ZMP	Minneapolis	MN	11
ZBW	Boston (Nashua, NH)	NH	11
ZAB	Albuquerque	NM	10
ZNY	New York	NY	12
ZOB	Cleveland (Oberlin)	OH	12
ZME	Memphis	TN	12
ZFW	Fort Worth	TX	12
ZHU	Houston	TX	11
ZLC	Salt Lake City	UT	10
ZDC	Washington (Leesburg)	VA	12
ZSE	Seattle (Auburn)	WA	10

Terminal Radar Approach Control

A TRACON is a standalone air traffic control facility providing radar control service to aircraft arriving or departing the primary airport and adjacent airports and to aircraft transiting the facility's airspace. Controllers in a stand-alone TRACON do not work in a control tower. Standalone TRACON's will always be identified with a letter followed by two numbers. A list of existing TRACON's is below:

TABLE 1-3 TERMINAL RADAR APPROACH CONTROL FACILITIES

Identification	Location	State	Level
A11	Anchorage	AK	8
A80	Atlanta (Peachtree City)	GA	12
A90	Boston (Merrimack, NH)	NH	11
C90	Chicago (Elgin)	IL	12
D01	Denver	CO	11
D10	Dallas/Ft Worth	TX	12
D21	Detroit	MI	11
E10	High Desert (Edwards AFB)	CA	8
F11	Central Florida (Orlando)	FL	11
I90	Houston	TX	12
K90	Cape (Falmouth, MA)	MA	7
L30	Las Vegas	NV	11
M03	Memphis	TN	9
M98	Minneapolis	MN	11
N90	New York (Westbury)	NY	12
NCT	Northern California (Mather)	CA	12
NMM	Meridian	MS	7
P31	Pensacola	FL	9
P50	Phoenix	AZ	11
P80	Portland	OR	9
PCT	Potomac	VA	12
R90	Omaha (Bellevue)	NE	8
S46	Seattle (Burien)	WA	10
S56	Salt Lake City	UT	10
SCT	Southern California (San Diego)	CA	12
T75	St Louis (St. Charles)	MO	10
U90	Tucson	AZ	8
Y90	Yankee (Windsor Locks)	CT	9

Different Control Tower Configurations

Several variations of air traffic control towers (ATCT) exist. These different types of towers are designed to handle different complexity and volume, as well as a different clientele—VFR or IFR, for example, or commercial versus general aviation traffic. Some of the different ATCT configurations are explained below.

Tower Without Radar

A control tower without radar is an air traffic control facility that provides airport air traffic control service using visual means only. These facilities are located at airports where the principal users are primarily low-performance aircraft operating VFR. These are commonly called "VFR towers". Most of these facilities are now operated by private companies under contract to the FAA.

Tower With Radar

A control tower with radar is an air traffic control facility that provides traffic advisories, spacing, sequencing, and separation services to VFR and IFR aircraft operating within the vicinity of the airport using a combination of radar and direct observations.

Combined Radar Approach Control and Tower With Radar

A Combined Radar Approach Control and Tower with Radar provides radar control services to aircraft arriving or departing the primary airport and adjacent airports and to aircraft transiting the facility's airspace, but it also has a tower to control traffic landing and departing at the primary airport. This facility would be operationally divided into two functional areas: radar approach control and tower. These two functional areas might be located within the same facility or in close proximity to one another, and controllers assigned to this facility would be certified and commonly work in both areas; this type of facility is commonly referred to as an "up-down." These are ideal training facilities, as developmental controllers receive both a tower and a radar controller rating.

Combined Nonradar Approach Control and Tower without Radar

This is an air traffic control facility that provides air traffic control services for the airport at which the tower is located and, without the use of radar, provides approach and departure control services to aircraft operating under IFR to and from one or more adjacent airports. There are very few of these facilities left in the domestic United States. Typically, the approach control function of a facility like this would be delegated to a nearby TRACON, and the tower would then become a VFR tower.

Below is a list of all current FAA towers and their classification.

TABLE 1-4 VARIOUS CONTROL TOWER FACILITIES

Ident.	Location	State	Level	Classification
ANC	Anchorage International	AK	8	Tower with Radar
FAI	Fairbanks Tower	AK	7	Combined TRACON & Tower with Radar
JNU	Juneau International	AK	5	Tower with Radar
MRI	Anchorage/Merrill Field	AK	7	Tower with Radar
BHM	Birmingham	AL	8	Combined TRACON & Tower with Radar
HSV	Huntsville Tower	AL	7	Combined TRACON & Tower with Radar
MGM	Montgomery Rgnl/Dannelly	AL	7	Combined TRACON & Tower with Radar
MOB	Mobile Regional	AL	8	Combined TRACON & Tower with Radar
FSM	Fort Smith Tower	AR	7	Combined TRACON & Tower with Radar
LIT	Little Rock Adams Field	AR	8	Combined TRACON & Tower with Radar
DVT	Phoenix-Deer Valley Muni	AZ	8	Tower with Radar
FFZ	Mesa/Falcon Field	AZ	7	Tower with Radar
GCN	Grand Canyon Municipal	AZ	5	Tower with Radar
PHX	Phoenix Sky Harbor Intl	AZ	11	Tower with Radar
PRC	Prescott/E A Love Field	AZ	7	Tower with Radar
SDL	Scottsdale	AZ	7	Tower with Radar
TUS	Tucson International	AZ	8	Tower with Radar
APC	Napa Tower	CA	4	Tower with Radar
BFL	Bakersfield/Meadows Fld	CA	7	Combined TRACON & Tower with Radar
BUR	Burbank-Glendale-Pasadena	CA	7	Tower with Radar
CCR	Concord/Buchanan Field	CA	5	Tower with Radar
CMA	Camarillo	CA	6	Tower with Radar
CNO	Chino	CA	7	Tower with Radar
CRQ	Carlsbad/McClellan Palomar	CA	7	Tower with Radar

Ident.	Location	State	Level	Classification
EMT	El Monte	CA	5	Tower with Radar
FAT	Fresno Yosemite Int`l	CA	8	Combined TRACON & Tower with Radar
HWD	Hayward Air Terminal	CA	6	Tower with Radar
LAX	Los Angeles International	CA	11	Tower with Radar
LGB	Long Beach/Daughtery Field	CA	8	Tower with Radar
LVK	Livermore Municipal	CA	6	Tower with Radar
MRY	Monterey Peninsula	CA	5	Tower with Radar
MYF	San Diego/Montgomery Field	CA	7	Tower with Radar
OAK	Oakland Tower	CA	7	Tower with Radar
ONT	Ontario International	CA	6	Tower with Radar
PAO	Palo Alto	CA	6	Tower with Radar
POC	La Verne/Brackett Field	CA	6	Tower with Radar
PSP	Palm Springs Regional	CA	7	Tower with Radar
RHV	San Jose/Reid-Hillview	CA	6	Tower with Radar
SAN	San Diego Int`l/Lindbergh	CA	7	Tower with Radar
SBA	Santa Barbara Municipal	CA	8	Combined TRACON & Tower with Radar
SCK	Stockton Metropolitan	CA	5	Tower with Radar
SEE	San Diego/Gillespie Field	CA	7	Tower with Radar
SFO	San Francisco Int`l	CA	9	Tower with Radar
SJC	San Jose Tower	CA	6	Tower with Radar
SMF	Sacramento International	CA	6	Tower with Radar
SMO	Santa Monica Municipal	CA	6	Tower with Radar
SNA	Santa Ana/John Wayne	CA	9	Tower with Radar
STS	Santa Rosa Sonoma County	CA	5	Tower with Radar
TOA	Torrance/Zamperini Field	CA	6	Tower with Radar
VNY	Van Nuys	CA	8	Tower with Radar
APA	Denver/Centennial	CO	8	Tower with Radar
ASE	Aspen Tower	CO	5	Combined TRACON & Tower with Radar
BJC	Denver/Jeffco	CO	6	Tower with Radar
COS	Colorado Springs Municipal	CO	8	Combined TRACON & Tower with Radar
DEN	Denver International	CO	12	Tower with Radar
PUB	Pueblo Memorial	CO	7	Tower with Radar
BDL	Windsor Locks/Bradley Intl	CT	7	Tower with Radar
DCA	Washington National	DC	10	Tower with Radar
ILG	Wilmington/New Castle Cnty	DE	6	Tower with Radar
DAB	Daytona Beach Int`l	FL	9	Combined TRACON & Tower with Radar
FLL	Ft Lauderdale/Hollywood	FL	9	Tower with Radar
FPR	Fort Pierce	FL	6	Tower with Radar
FXE	Fort Lauderdale Executive	FL	7	Tower with Radar

Ident.	Location	State	Level	Classification
JAX	Jacksonville Int`l	FL	9	Combined TRACON & Tower with Radar
MCO	Orlando International	FL	11	Tower with Radar
MIA	Miami International	FL	12	Combined TRACON & Tower with Radar
ORL	Orlando Executive	FL	5	Tower with Radar
PBI	Palm Beach International	FL	9	Combined TRACON & Tower with Radar
PIE	St Petersburg Clearwater	FL	7	Tower with Radar
PNS	Pensacola Regional	FL	6	Tower with Radar
RSW	Ft Myers/SW FL Int`l	FL	8	Combined TRACON & Tower with Radar
SFB	Orlando/Sanford	FL	8	Tower with Radar
SRQ	Sarasota Bradenton	FL	6	Tower with Radar
TLH	Tallahassee Regional	FL	7	Combined TRACON & Tower with Radar
TMB	Miami/Kendall-Tamiami Exec	FL	7	Tower with Radar
TPA	Tampa International	FL	10	Combined TRACON & Tower with Radar
VRB	Vero Beach	FL	6	Tower with Radar
AGS	Augusta/Bush Field	GA	6	Combined TRACON & Tower with Radar
ATL	Hartsfield-Jackson Atlanta Int`l	GA	12	Tower with Radar
CSG	Columbus Metropolitan	GA	4	Tower with Radar
PDK	Atlanta/Dekalb-Peachtree	GA	7	Tower with Radar
SAV	Savannah International	GA	8	Combined TRACON & Tower with Radar
HNL	Honolulu International	HI	11	Tower with Radar
ITO	Hilo International	HI	7	Combined TRACON & Tower with Radar
OGG	Maui/Kahului	HI	7	Tower with Radar
ALO	Waterloo Municipal	IA	5	Combined TRACON & Tower with Radar
CID	Cedar Rapids	IA	6	Combined TRACON & Tower with Radar
DSM	Des Moines International	IA	7	Combined TRACON & Tower with Radar
SUX	Sioux City/Sioux Gateway	IA	5	Combined TRACON & Tower with Radar
BOI	Boise Air Terminal	ID	8	Combined TRACON & Tower with Radar
TWF	Twin Falls	ID	5	Combined Non-Radar Appr Cntrl & Tower
ARR	Chicago/Aurora Municipal	IL	5	Tower with Radar
CMI	Champaign/Univ of Illinois	IL	7	Combined TRACON & Tower with Radar
CPS	Cahokia/St Louis Downtown	IL	6	Tower with Radar
DPA	Chicago/Du Page	IL	5	Tower with Radar
MDW	Chicago Midway	IL	8	Tower with Radar
MLI	Moline/Quad City Int`l	IL	6	Combined TRACON & Tower with Radar
ORD	Chicago/O`Hare Int`l	IL	12	Tower with Radar
PIA	Greater Peoria Regional	IL	6	Combined TRACON & Tower with Radar
PWK	Chicago/Palwaukee Muni	IL	6	Tower with Radar
RFD	Rockford	IL	7	Combined TRACON & Tower with Radar
SPI	Springfield/Capital	IL	6	Combined TRACON & Tower with Radar
EVV	Evansville Regional	IN	6	Combined TRACON & Tower with Radar
FWA	Fort Wayne International	IN	7	Combined TRACON & Tower with Radar

Ident.	Location	State	Level	Classification
HUF	Terre Haute/Hulman Rgnl	IN	5	Combined TRACON & Tower with Radar
IND	Indianapolis International	IN	9	Combined TRACON & Tower with Radar
LAF	Lafayette/Purdue U	IN	4	Tower without Radar
SBN	South Bend/MI Rgnl Trans	IN	7	Combined TRACON & Tower with Radar
ICT	Wichita Mid Continent	KS	9	Combined TRACON & Tower with Radar
CVG	Covington/Cincinnati Int`l	KY	11	Combined TRACON & Tower with Radar
LEX	Lexington/Blue Grass	KY	7	Combined TRACON & Tower with Radar
LOU	Louisville Bowman	KY	5	Tower with Radar
SDF	Louisville Intl/Standiford	KY	9	Combined TRACON & Tower with Radar
BTR	Baton Rouge Ryan Field	LA	7	Combined TRACON & Tower with Radar
LCH	Lake Charles	LA	6	Combined TRACON & Tower with Radar
LFT	Lafayette	LA	7	Combined TRACON & Tower with Radar
MLU	Monroe Regional	LA	6	Combined TRACON & Tower with Radar
MSY	New Orleans Int`l/Moisant	LA	9	Combined TRACON & Tower with Radar
NEW	New Orleans/Lakefront	LA	5	Tower with Radar
SHV	Shreveport Regional	LA	7	Combined TRACON & Tower with Radar
ACK	Nantucket Memorial	MA	7	Tower with Radar
ACK	Nantucket Memorial	MA	7	Tower with Radar
BED	Bedford/Hanscom Field	MA	7	Tower with Radar
BOS	Boston/Logan International	MA	10	Tower with Radar
ADW	Camp Springs/Andrews AFB	MD	6	Tower with Radar
BWI	Baltimore-Washington Int`l	MD	9	Tower with Radar
BGR	Bangor Tower	ME	5	Combined TRACON & Tower with Radar
PWM	Portland Int`l Jetport	ME	7	Combined TRACON & Tower with Radar
ARB	Ann Arbor Municipal	MI	5	Tower with Radar
AZO	Kalamazoo/Battle Creek Int	MI	6	Combined TRACON & Tower with Radar
DTW	Detroit Metro Wayne Co	MI	11	Tower with Radar
FNT	Flint Tower	MI	5	Combined TRACON & Tower with Radar
GRR	Grand Rapids/Kent Co Int`l	MI	8	Combined TRACON & Tower with Radar
LAN	Lansing/Capital City	MI	8	Combined TRACON & Tower with Radar
MBS	Saginaw/MBS International	MI	6	Combined TRACON & Tower with Radar
MKG	Muskegon County	MI	6	Combined TRACON & Tower with Radar
PTK	Pontiac/Oakland Cnty Int`l	MI	7	Tower with Radar
TVC	Traverse City	MI	5	Tower with Radar
YIP	Detroit Willow Run	MI	5	Tower with Radar
DLH	Duluth International	MN	6	Combined TRACON & Tower with Radar
FCM	Flying Cloud Tower	MN	5	Tower with Radar

Ident.	Location	State	Level	Classification
FCM	Flying Cloud Tower	MN	5	Tower with Radar
MIC	Minneapolis/Crystal	MN	4	Tower with Radar
MIC	Minneapolis/Crystal	MN	4	Tower with Radar
MSP	Minneapolis-St Paul Int`l	MN	11	Tower with Radar
RST	Rochester International	MN	6	Combined TRACON & Tower with Radar
STP	St Paul Downtown	MN	6	Tower with Radar
MCI	Kansas City International	MO	9	Combined TRACON & Tower with Radar
MKC	Kansas City Downtown	MO	5	Tower with Radar
SGF	Springfield-Branson Rgnl	MO	7	Combined TRACON & Tower with Radar
STL	Lambert-St Louis Int`l	MO	9	Tower with Radar
SUS	Spirit of St Louis	MO	5	Tower with Radar
GPT	Gulfport/Biloxi Regional	MS	7	Combined TRACON & Tower with Radar
JAN	Jackson International	MS	7	Combined TRACON & Tower with Radar
BIL	Billings Logan Int`l	MT	7	Combined TRACON & Tower with Radar
GTF	Great Falls International	MT	5	Combined TRACON & Tower with Radar
HLN	Helena Tower	MT	5	Combined Non-Radar Appr Cntrl & Tower
AVL	Asheville Regional	NC	6	Combined TRACON & Tower with Radar
CLT	Charlotte/Douglas Int`l	NC	12	Combined TRACON & Tower with Radar
FAY	Fayetteville Regional	NC	7	Combined TRACON & Tower with Radar
GSO	Greensboro/Piedmont Triad	NC	8	Combined TRACON & Tower with Radar
ILM	Wilmington/New Hanover Int	NC	7	Combined TRACON & Tower with Radar
RDU	Raleigh-Durham Int`l	NC	9	Combined TRACON & Tower with Radar
BIS	Bismarck Municipal	ND	5	Combined TRACON & Tower with Radar
FAR	Fargo/Hector International	ND	6	Combined TRACON & Tower with Radar
GFK	Grand Forks Tower	ND	9	Tower with Radar
LNK	Lincoln Municipal	NE	7	Tower with Radar
OMA	Omaha	NE	6	Tower with Radar
MHT	Manchester	NH	5	Tower with Radar
ACY	Atlantic City Int`l	NJ	8	Combined TRACON & Tower with Radar
CDW	Caldwell/Essex County	NJ	5	Tower with Radar
EWR	Newark Tower	NJ	11	Tower with Radar
MMU	Morristown Tower	NJ	5	Tower with Radar
TEB	Teterboro	NJ	7	Tower with Radar
ABQ	Albuquerque International	NM	9	Combined TRACON & Tower with Radar
ROW	Roswell Industrial Air Cnt	NM	7	Combined TRACON & Tower with Radar
LAS	Las Vegas/Mc Carran Int`l	NV	11	Tower with Radar
RNO	Reno/Tahoe International	NV	5	Tower with Radar
VGT	North Las Vegas	NV	6	Tower with Radar
VGT	North Las Vegas	NV	6	Tower with Radar
ALB	Albany County	NY	8	Combined TRACON & Tower with Radar
BGM	Binghamton Rgnl/Link Field	NY	5	Combined TRACON & Tower with Radar

Ident.	Location	State	Level	Classification
BUF	Greater Buffalo Int`l	NY	8	Combined TRACON & Tower with Radar
ELM	Elmira/Corning Regional	NY	5	Combined TRACON & Tower with Radar
FRG	Farmingdale/Republic	NY	7	Tower with Radar
HPN	White Plains/Westchester	NY	7	Tower with Radar
ISP	Islip/Long Isl. MacArthur	NY	7	Tower with Radar
JFK	Kennedy Tower	NY	11	Tower with Radar
LGA	La Guardia	NY	10	Tower with Radar
POU	Poughkeepsie/Dutchess Co	NY	5	Tower with Radar
ROC	Greater Rochester Int`l	NY	7	Combined TRACON & Tower with Radar
SYR	Syracuse Hancock Int`l	NY	8	Combined TRACON & Tower with Radar
CAK	Akron Canton Regional	OH	7	Combined TRACON & Tower with Radar
CLE	Cleveland Hopkins Int`l	OH	10	Combined TRACON & Tower with Radar
CMH	Port Columbus Int`l	OH	9	Combined TRACON & Tower with Radar
DAY	Dayton International	OH	5	Tower with Radar
MFD	Mansfield Lahm Municipal	OH	5	Combined TRACON & Tower with Radar
TOL	Toledo Express	OH	7	Combined TRACON & Tower with Radar
YNG	Youngstown Tower	OH	5	Combined TRACON & Tower with Radar
OKC	Oklahoma City/Will Rogers	OK	8	Combined TRACON & Tower with Radar
RVS	Tulsa/Riverside	OK	8	Tower with Radar
TUL	Tulsa International	OK	8	Combined TRACON & Tower with Radar
EUG	Eugene/M Sweet Field	OR	7	Combined TRACON & Tower with Radar
HIO	Portland-Hillsboro	OR	7	Tower with Radar
PDX	Portland International	OR	8	Tower with Radar
ABE	Allentown/Lehigh Valley	PA	7	Combined TRACON & Tower with Radar
AGC	Pittsburgh/Allegheny Cnty	PA	5	Tower with Radar
AVP	Wilkes-Barre Tower	PA	6	Combined TRACON & Tower with Radar
ERI	Erie International	PA	5	Combined TRACON & Tower with Radar
MDT	Harrisburg International	PA	8	Combined TRACON & Tower with Radar
PHL	Philadelphia International	PA	12	Combined TRACON & Tower with Radar
PIT	Pittsburgh International	PA	10	Combined TRACON & Tower with Radar
PNE	Northeast Philadelphia	PA	5	Tower with Radar
RDG	Reading Regional	PA	7	Combined TRACON & Tower with Radar
SJU	San Juan International	PR	7	Tower with Radar
PVD	Providence	RI	8	Combined TRACON & Tower with Radar
CAE	Columbia Metropolitan	SC	7	Combined TRACON & Tower with Radar
CHS	Charleston AFB/Int`l	SC	8	Combined TRACON & Tower with Radar
FLO	Florence City	SC	6	Combined TRACON & Tower with Radar
GSP	Greenville-Spartanburg	SC	7	Combined TRACON & Tower with Radar
MYR	Myrtle Beach International	SC	7	Combined TRACON & Tower with Radar
FSD	Sioux Falls/Foss Field	SD	6	Combined TRACON & Tower with Radar
BNA	Nashville International	TN	9	Combined TRACON & Tower with Radar
CHA	Chattanooga/Lovell Field	TN	7	Combined TRACON & Tower with Radar
MEM	Memphis Tower	TN	9	Tower with Radar

Ident.	Location	State	Level	Classification
TRI	Tri-Cities Regional	TN	6	Combined TRACON & Tower with Radar
TYS	Knoxville/McGhee Tyson	TN	7	Combined TRACON & Tower with Radar
ABI	Abilene Tower	TX	6	Combined TRACON & Tower with Radar
ACT	Waco Tower	TX	5	Combined TRACON & Tower with Radar
ADS	Dallas Addison	TX	6	Tower with Radar
ADS	Dallas Addison	TX	6	Tower with Radar
AFW	Ft Worth/Alliance	TX	5	Tower with Radar
AMA	Amarillo Tower	TX	6	Combined TRACON & Tower with Radar
AUS	Austin Tower	TX	8	Combined TRACON & Tower with Radar
BPT	Beaumont Port Arthur	TX	6	Tower with Radar
CRP	Corpus Christi	TX	9	Combined TRACON & Tower with Radar
DAL	Dallas Love Field	TX	8	Tower with Radar
DFW	Dallas/Ft Worth Int`l	TX	12	Tower with Radar
DWH	Tomball D W Hooks	TX	7	Tower with Radar
ELP	El Paso International	TX	7	Combined TRACON & Tower with Radar
FTW	Fort Worth Meacham	TX	5	Tower with Radar
GGG	Longview	TX	7	Combined TRACON & Tower with Radar
HOU	Houston Hobby	TX	8	Tower with Radar
IAH	Houston/G Bush Intercont`l	TX	12	Tower with Radar
LBB	Lubbock International	TX	7	Combined TRACON & Tower with Radar
MAF	Midland International	TX	8	Combined TRACON & Tower with Radar
SAT	San Antonio International	TX	10	Combined TRACON & Tower with Radar
SLC	Salt Lake City Int`l	UT	10	Tower with Radar
HEF	Manassas Rgnl/Davis Fld	VA	5	Tower with Radar
IAD	Washington Dulles Int`l	VA	11	Tower with Radar
ORF	Norfolk International	VA	9	Combined TRACON & Tower with Radar
PHF	Newport News/P Henry Int`l	VA	7	Tower with Radar
RIC	Richmond International	VA	6	Tower with Radar
ROA	Roanoke Regional	VA	7	Combined TRACON & Tower with Radar
STT	St Thomas H S Truman	VI	5	Tower with Radar
BTV	Burlington Tower	VT	6	Combined TRACON & Tower with Radar
BFI	Boeing Tower	WA	7	Tower with Radar
GEG	Spokane International	WA	8	Combined TRACON & Tower with Radar
MWH	Moses Lake/Grant Co Int`l	WA	6	Combined TRACON & Tower with Radar
PAE	Everett Paine Field	WA	6	Tower with Radar
PSC	Pasco Tri Cities	WA	7	Combined TRACON & Tower with Radar
SEA	Seattle Tacoma Int`l	WA	9	Tower with Radar
GRB	Green Bay Tower	WI	6	Combined TRACON & Tower with Radar
MKE	Milwaukee/Gen Mitchell Int	WI	9	Combined TRACON & Tower with Radar
MSN	Madison/Dane Cnty Regional	WI	8	Combined TRACON & Tower with Radar
CKB	Clarksburg/Benedum	WV	5	Combined TRACON & Tower with Radar
CRW	Charleston/Yeager	WV	7	Combined TRACON & Tower with Radar
HTS	Huntington	WV	6	Combined TRACON & Tower with Radar

Ident.	Location	State	Level	Classification
CPR	Casper	WY	5	Combined TRACON & Tower with Radar

Combined Control Facility (CERAP)

A CERAP provides approach control services for one or more airports as well as enroute air traffic control for a large area of airspace. Some may provide tower services along with approach control and enroute services. This is an uncommon facility found primarily in remote or off-shore areas.

TABLE 1-5 COMBINED CONTROL FACILITIES

Identification	Location	State	Level
HCF	Honolulu Control Facility	HI	11
ZSU	San Juan	PR	9
ZUA	Guam	GU	8

Federal Contract Air Traffic Control Towers

In 1982, Congress authorized the FAA to initiate a pilot program to contract out air traffic control services for five VFR towers that had closed as a result of the controllers strike of 1981. Since then, the contract tower program has expanded to over 200 contract control towers.

Contract controllers providing air traffic control services in towers in the contract tower program must meet the same controller certification requirements as federal controllers and are certified by the FAA. Contract towers do not usually hire untrained controllers. Instead, they rely on retired FAA and/or military controllers for their staffing needs. The supply of these controllers is diminishing, however, and there is concern as to how these control facilities might be staffed in the future.

TABLE 1-6 FEDERAL CONTRACT AIR TRAFFIC CONTROL TOWERS

Identification	Location	State
ADQ	Kodiak	AK
AKN	King Salmon Tower	AK
BET	Bethel	AK
ENA	Kenai Municipal	AK
BFM	Mobile Downtown	AL
DHN	Dothan	AL
TCL	Tuscaloosa Municipal	AL
ASG	Springdale Municipal	AR
FYV	Fayetteville/Drake Field	AR
ROG	Rogers	AR
TXK	Texarkana Regional	AR
XNA	Northwest Arkansas Tower	AR
CHD	Chandler Municipal	AZ
FLG	Flagstaff-Pulliam	AZ
GEU	Glendale Municipal	AZ
GYR	Phoenix-Goodyear Municipal	AZ
IFP	Laughlin/Bullhead Int'l	AZ
IWA	Phoenix/Williams Gateway	AZ
RYN	Tucson/Ryan Field	AZ
CIC	Chico	CA

Identification	Location	State
FUL	Fullerton Municipal	CA
HHR	Hawthorne Mun/Northrop Fld	CA
MER	Atwater/Castle AFB	CA
MHR	Sacramento/Mather	CA
MOD	Modesto/City-County	CA
OXR	Oxnard	CA
PMD	Palmdale	CA
RAL	Riverside Municipal	CA
RDD	Redding	CA
RNM	Ramona Airport	CA
SAC	Sacramento Executive	CA
SBP	San Luis Obispo	CA
SDM	San Diego/Brown Fld Muni	CA
SMX	Santa Maria/Hancock Field	CA
SNS	Salinas Municipal	CA
SQL	San Carlos	CA
VCV	Victorville Tower	CA
WHP	Los Angeles/Whiteman	CA
WJF	Lancaster/Gen Fox Airfield	CA
EGE	Eagle County Regional	CO
FTG	Front Range /Denver	CO
GJT	Grand Junction/Walker Fld	CO
BDR	Bridgeport/Sikorsky Mem	CT
DXR	Danbury Municipal	CT
GON	Groton-New London	CT
HFD	Hartford-Brainard	CT
HVN	New Haven/Tweed	CT
OXC	Waterbury-Oxford	CT
APF	Naples Municipal	FL
BCT	Boca Raton	FL
BKV	Brooksville Tower	FL
CRG	Jacksonville/Craig Muni	FL
ECP	Panama City Tower	FL
EVB	New Smyrna Beach Municipal	FL
EYW	Key West International	FL
FMY	Fort Myers/Page Field	FL
GNV	Gainesville Regional	FL
HWO	Hollywood/North Perry	FL
ISM	Orlando/Kissimmee Muni	FL
LAL	Lakeland/Linder Regional	FL
LEE	Leesburg Municipal	FL
MLB	Melbourne International	FL
OCF	Ocala Airport	FL
OMN	Ormond Beach Municipal	FL
OPF	Miami/Opa Locka	FL
PGD	Punta Gorda	FL
PMP	Pompano Beach Airpark	FL
SGJ	St Augustine	FL

Identification	Location	State
SPG	St Petersburg	FL
SUA	Stuart/Witham Field	FL
TIX	Titusville/Space Coast Rgn	FL
VQQ	Jacksonville/Cecil Field	FL
XFL	Flagler Tower	FL
ABY	Albany/SW GA Regional	GA
AHN	Athens/Ben Epps	GA
FTY	Atlanta/Fulton County	GA
LZU	Lawrenceville/Gwinnett Cty	GA
MCN	Macon/Middle GA Regional	GA
RYY	Marietta/Cobb County	GA
GUM	Agana/Guam International	GU
JRF	Kalaeloa Arpt	HI
KOA	Kailua/Kona International	HI
LIH	Lihue	HI
MKK	Kaunakakai/Molokai	HI
DBQ	Dubuque Regional	IA
IDA	Idaho Falls/Fanning Field	ID
LWS	Lewiston/Nez Perce Cnty	ID
PIH	Pocatello Regional	ID
SUN	Hailey/Friedman Memorial	ID
ALN	Alton/St Louis Regional	IL
BMI	Bloomington/Central IL Rgn	IL
DEC	Decatur	IL
MDH	Carbondale/Southern IL	IL
MWA	Marion/Williamson Cty Rgnl	IL
UGN	Chicago/Waukegan Regional	IL
BAK	Columbus Municipal	IN
BMG	Bloomington/Monroe Cnty	IN
GYY	Gary Regional	IN
MIE	Muncie/Delaware County	IN
FOE	Topeka/Forbes Field	KS
GCK	Garden City	KS
HUT	Hutchinson Municipal	KS
IXD	Olathe/New Century Aircenter	KS
MHK	Manhattan	KS
OJC	Olathe/Johnson Cnty Exec	KS
SLN	Salina Municipal	KS
TOP	Topeka/P Billard Municipal	KS
OWB	Owensboro/Daviess Cnty	KY
PAH	Paducah/Barkley Regional	KY
AEX	Alexandria International	LA
ARA	New Iberia/Acadiana Rgnl	LA
CWF	Lake Charles/Chennault	LA
DTN	Shreveport Downtown	LA
HUM	Houma-Terrebonne	LA
BAF	Westfield/Barnes Municipal	MA
BVY	Beverly Municipal	MA

Identification	Location	State
EWB	New Bedford Regional	MA
HYA	Hyannis	MA
LWM	Lawrence Municipal	MA
MVY	Martha's Vineyard	MA
ORH	Worcester Regional	MA
OWD	Norwood Memorial	MA
ESN	Easton/Newnam Field	MD
FDK	Frederick Municipal	MD
HGR	Hagerstown/WA Cnty Rgnl	MD
MTN	Baltimore/Martin State	MD
SBY	Salisbury/OC Wicomico Rgnl	MD
BTL	Battle Creek/Kellogg	MI
DET	Detroit City	MI
JXN	Jackson Cnty-Reynolds Fld	MI
SAW	Marquette Sawyer AFB	MI
ANE	Minneapolis/Anoka Cnty	MN
STC	St Cloud Municipal	MN
BBG	Branson Tower	MO
COU	Columbia Regional	MO
JEF	Jefferson City Memorial	MO
JLN	Joplin Regional	MO
STJ	St Joseph/Rosecrans Mem	MO
GSN	Saipan International	MP
GLH	Greenville/Mid Delta Rgnl	MS
GTR	Golden Triangle Regional	MS
HKS	Jackson/Hawkins Field	MS
HSA	Bay St Louis/Stennis Intl	MS
MEI	Meridian/Key Field	MS
OLV	Olive Branch	MS
TUP	Tupelo Regional	MS
BZN	Bozeman/Gallatin Field	MT
GPI	Glacier Park Intl (FCA-Kalispell)	MT
MSO	Missoula International	MT
EWN	New Bern/Craven Co Rgnl	NC
HKY	Hickory Regional	NC
INT	Winston Salem	NC
ISO	Kinston Regional	NC
JQF	Concord Regional	NC
MOT	Minot International	ND
GRI	Grand Isl./Central NE Rgnl	NE
ASH	Nashua/Boire Field	NH
LEB	Lebanon Municipal	NH
TTN	Trenton Mercer	NJ
AEG	Double Eagle II	NM
FMN	Farmington/4 Corners Rgnl	NM
HOB	Hobbs/Lea County	NM
SAF	Santa Fe Municipal	NM
HND	Henderson ATCT	NV

Identification	Location	State
FOK	Suffolk Tower	NY
IAG	Niagara Falls Int'l	NY
ITH	Ithaca/Tompkins County	NY
RME	Rome	NY
SWF	Newburgh/Stewart Int`l	NY
BKL	Lakefront Tower	OH
CGF	Cleveland Cuyahoga County	OH
LUK	Cincinnati/Lunken Field	OH
OSU	Columbus/OH State U	OH
TZR	Columbus/Bolton Field	OH
ADM	Ardmore Municipal	OK
LAW	Lawton Municipal	OK
OUN	Norman/U of OK Westheimer	OK
PWA	Oklahoma City/Wiley Post	OK
SWO	Stillwater Municipal	OK
WDG	Enid Woodring Municipal	OK
LMT	Klamath Falls Int'l	OR
MFR	Medford/Rogue Valley Int'l	OR
OTH	Southwest Oregon Regional	OR
PDT	Pendleton/Eastern OR Rgnl	OR
RDM	Redmond/Roberts Field	OR
SLE	Salem/Mc Nary Field	OR
TTD	Portland-Troutdale	OR
CXY	Harrisburg/Capital City	PA
IPT	Williamsport Regional	PA
LBE	Latrobe	PA
LNS	Lancaster	PA
UNV	University Park Airport	PA
BQN	Ramey AFB	PR
SIG	San Juan/F L R Dominicci	PR
CRE	North Myrtle Beach	SC
GMU	Greenville Downtown	SC
GYH	Greenville/Donaldson	SC
HXD	Hilton Head Island	SC
RAP	Rapid City Regional	SD
MKL	Jackson/McKellar	TN
MQY	Smyrna	TN
NQA	Millington Municipal	TN
BAZ	New Braunfels Municipal	TX
BRO	Brownsville/S Padre Island	TX
CLL	College Station Tower	TX
CNW	Waco James Connally	TX
CXO	Conroe	TX
DTO	Denton	TX
FWS	Fort Worth Spinks	TX
GKY	Arlington Municipal	TX
GLS	Galveston	TX
GPM	Grand Prairie Tower	TX

Identification	Location	State
GTU	Georgetown	TX
HQZ	Mesquite/P L Hudson Muni	TX
HRL	Harlingen/Valley Int'l	TX
HYI	San Marcos	TX
LRD	Laredo International	TX
MFE	Mc Allen/Miller Int'l	TX
RBD	Dallas Redbird	TX
SGR	Sugarland	TX
SJT	San Angelo/Mathis Field	TX
SSF	San Antonio/Stinson Field	TX
TKI	McKinney Municipal	TX
TYR	Tyler Pounds Field	TX
VCT	Victoria Regional	TX
OGD	Ogden/Hinckley	UT
PVU	Provo Municipal	UT
CHO	Charlottesville Albemarle	VA
LYH	Lynchburg Regional	VA
STX	Christiansted (St Croix)	VI
ALW	Walla Walla Regional	WA
BLI	Bellingham International	WA
OLM	Olympia	WA
RNT	Renton Municipal	WA
SFF	Spokane/Felts Field	WA
TIW	Tacoma Narrows	WA
YKM	Yakima Air Terminal	WA
ATW	Appleton/Outagamie County	WI
CWA	Mosinee/Central WI	WI
EAU	Eau Claire	WI
ENW	Kenosha Regional	WI
JVL	Janesville/Rock County	WI
LSE	La Crosse Municipal	WI
MWC	Milwaukee/LJ Timmerman	WI
OSH	Oshkosh/Wittman Regional	WI
UES	Waukesha County	WI
HLG	Wheeling/OH County	WV
LWB	Greenbrier Tower	WV
MGW	Morgantown Municipal	WV
PKB	Parkersburg/Wood County	WV
CYS	Cheyenne	WY
JAC	Jackson Hole	WY

Flight Service Stations

Flight Service Stations (FSS) primarily provide preflight, in-flight, and enroute communications and weather services to private and corporate aircraft. They also coordinate search and rescue operations and provide operational support to air shows, conventions, and other aviation events. They are also the main facility for initiating and tracking Notices to Airmen (NOTAM's)

Flight Services Stations were previously owned and operated by the FAA. Due to the inherent limitations of radio communications in the early twentieth century, FSS stations were initially placed along major air routes spaced every 30 to 50 miles. As air travel grew, this eventually resulted in hundreds of stations across the country.

In 2005, the FAA awarded a private contract for the operation and staffing of AFSSs in the continental United States, Puerto Rico, and Hawaii to the Lockheed-Martin Corporation (LMC); flight service stations in Alaska are still operated by the FAA. Lockheed Martin assumed responsibility for providing flight services at these stations beginning in October 2005. The FAA still provides oversight, but LMC has the operational authority to deliver all FSS services. These services are provided to pilots through a system of FSS hubs located in Virginia, Arizona, and Texas.

All flight service specialists employed by LMC complete FAA-approved air traffic control training and are certified by the National Weather Service (NWS) as pilot weather briefers. AFSS employees who work for Lockheed Martin work in facilities similar to FAA operated air traffic control facilities and are considered to be part of the national airspace system. Individuals seeking careers in flight service should apply directly to Lockheed-Martin.

Chapter 2

Navigation, Airports, and the National Airspace System

Visual Navigation

In the early years of aviation, pilots were limited to daylight flying during good weather conditions. They used outside visual references to control their aircraft's attitude, relying on the natural horizon as a reference. They would note any changes in the flight attitude of their aircraft and make necessary control adjustments to their aircraft in level flight.

Pilots navigated from airport to airport using either pilotage or deduced reckoning (commonly called dead reckoning). Pilotage requires pilots to use a map of the surrounding area as a reference. They draw a line on the map, extending from the departure to the destination airport, and note any prominent landmarks to be passed in flight. As the aircraft passed these landmarks, the pilot notes any deviation from the planned flight path and adjusts the aircraft's heading to return to course.

Dead reckoning is a process whereby pilots take into account wind direction and speed, then calculate and plot a course to fly. Winds at the aircraft's cruising altitude usually cause the aircraft to drift either left or right of course. The calculated change in heading to counteract the crosswind component of the wind is known as the crosswind correction angle or wind correction angle. The resultant path the aircraft then flies over the ground is known as the ground track or the course. Most pilots flying VFR use a combination of pilotage and dead reckoning

Aeronautical Charts

Maps used by pilots in the early 1920s were common road maps available at automobile service stations. These maps were unsuitable for aerial navigation since they lacked the necessary landmark information needed to accurately navigate from one airport to the next. The U.S. government began to print air navigation charts, known as sectional charts around that time.

Sectional Charts

Sectional charts are aeronautical charts scaled 1:500,000 or about 8 statute miles to the inch. Sectional charts are still used today and depict the relevant information needed by pilots to navigate accurately and safely. This information includes cities, highways, railroads, airport locations, terrain features, and distinctive objects. Sectional charts also depict navigation aids, federal airways, and air traffic control facilities.

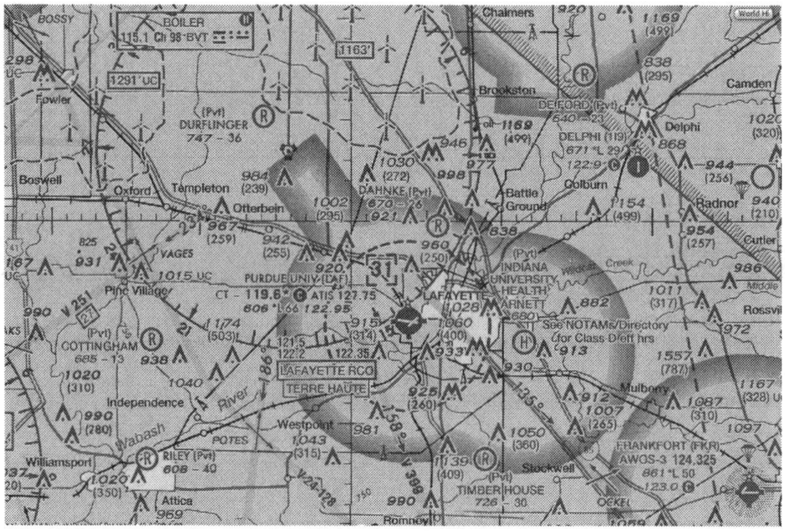

Figure 2-1. Sectional chart excerpt

Some pilots carry world aeronautical charts (WACs) instead of sectionals during IFR flights. (See figure 2-2). WACs are similar to sectionals but are scaled 1:1,000,000 or about 16 miles to the inch. They present less-detailed information to the pilot but cover a larger area than a sectional chart.

Instrument Flying

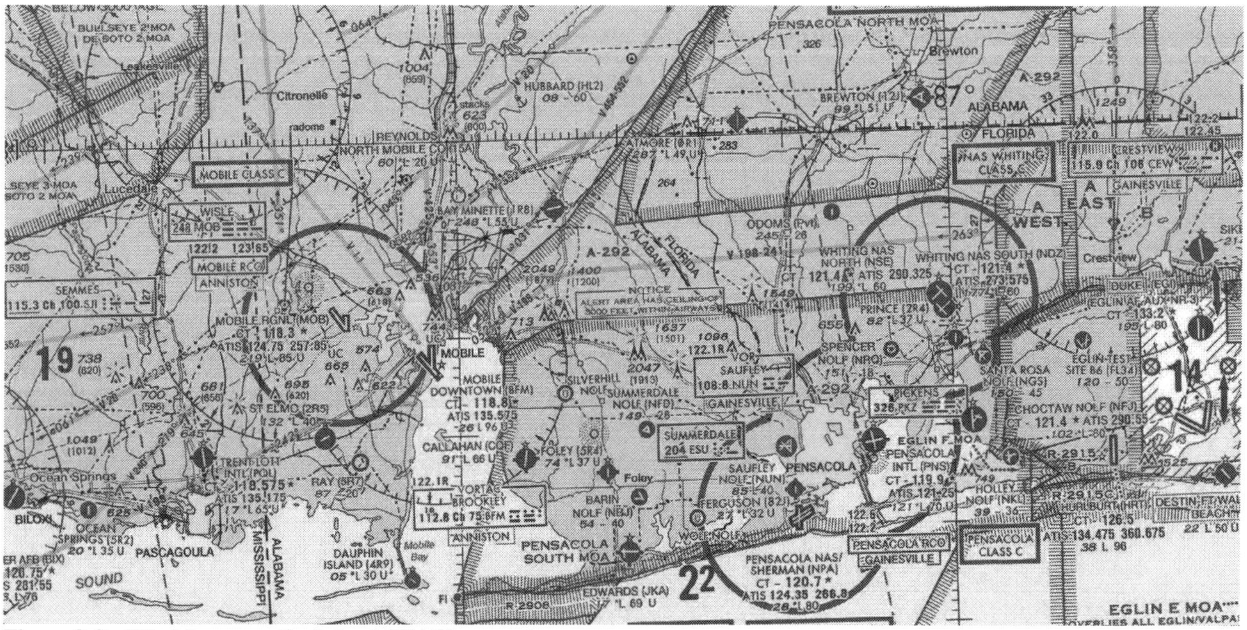

Figure 2-2. World aeronautical chart (WAC) excerpt

Pilots flying using instrument flight rules (IFR) have more demanding navigational needs. Depending on their cruising altitude, IFR pilots use either low altitude (below 18,000') or high altitude (18,000' and above) enroute charts to navigate between airports. These charts are similar to sectionals and WACs; however, since IFR pilots are not navigating with visual references, these charts omit most terrain and other ground-based information used by VFR pilots. IFR charts provide more electronic navigational information as well as radio frequencies and cruising altitude information. (See figures 2-3 and 2-4). An excellent guide to understanding both VFR and IFR charts can be downloaded from the FAA publications website; the document is entitled "Aeronautical Chart Users Guide".

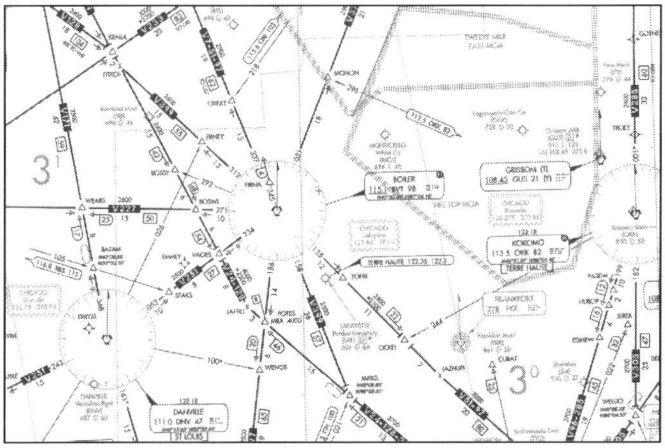

Figure 2-3. Low altitudes IFR chart

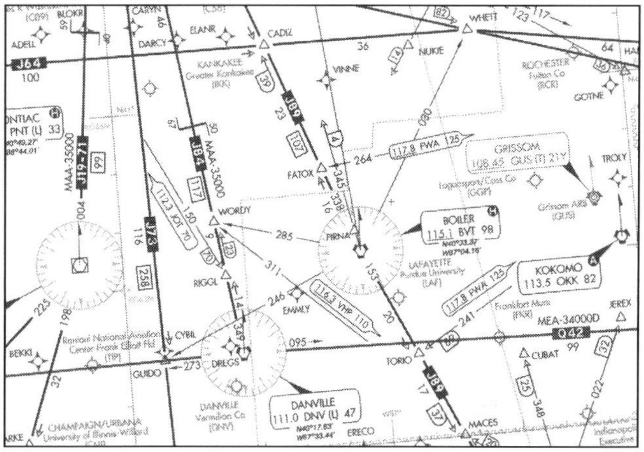

Figure 2-4. High altitude IFR chart

Navigation Systems

Pilots navigating using IFR charts and procedures utilize a number of navigation aids. Some are designed to be used strictly for enroute navigation; others are designed to guide aircraft to the runway for landing. Some navigation aids can be used for both purposes.

VHF Omnidirectional Range (VOR)

One of the first modern electronic navigation systems was the VOR. The VOR was developed by the predecessor of the FAA as the US standard navigation aid in 1946 and was then selected as the international civil navigation standard by the International Civil Aviation Organization (ICAO) in 1949. It is still used world-wide. VOR's are the basis for the victor, or lower-level, airway system in the United States.

Figure 2-5. VORTAC ground station

VORs are assigned a frequency between 108.10 and 117.90 MHz. When operating, the aircraft's VOR receiver measures the phase difference between two signals transmitted by the VOR to determine the direction of the aircraft in relation to the transmitter.

If an aircraft, for example, is located directly east of a VOR, it is said to be on the 090° radial of the VOR. An aircraft directly south of the VOR will on the 180° radial, west is the 270° radial and north is the 360° degree radial. There are potentially 360 different courses (radials) that the pilot can select from.

To use the VOR, the radial to be flown is selected by the pilot (see Figure 2–6) using the omni bearing selector

(OBS) knob. The indicator on the instrument panel informs the pilot whether the selected course leads to the station or away from it (known as the To–From flag). The VOR indicator also displays any lateral deviation from the selected course, using a vertical pointer known as the course deviation indicator (CDI). If the aircraft is to the right of the selected course, the CDI needle will be to the left of center, advising the pilot to alter the aircraft's course to the left. If the aircraft is left of course, the CDI will be right of center. If the aircraft is precisely located on the radial selected by the pilot, the CDI will be centered.

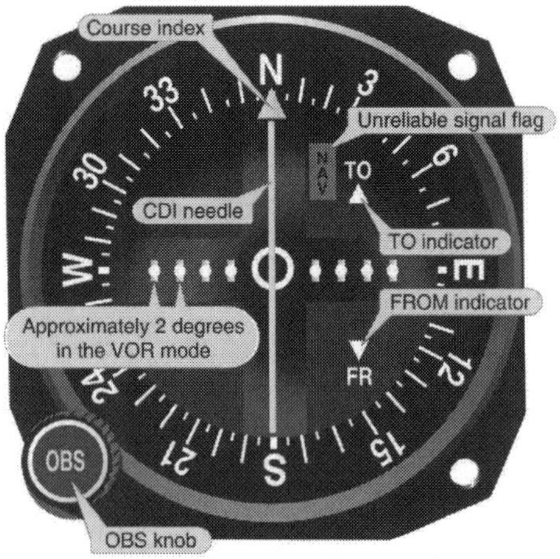

Figure 2-6. VOR indicator

VOR Categories

VHF radio transmissions are line of sight, so low-flying aircraft might be unable to receive the VOR signal if they are "below the horizon" of the VOR transmitter. This forces the FAA to place VORs no farther than 80 miles from each other to ensure adequate reception for aircraft operating at low altitudes. Since there are only a limited number of frequencies that can be assigned to the VORs across the country, some have to be assigned the same frequency. This can cause interference problems for aircraft operating at high altitudes, as they could receive signals from two or more VORs operating on the same frequency with the resulting interference rendering the navigation signal unusable.

The FAA solved this problem by designating each VOR as either a terminal, low, or high-altitude VOR. Due to its "line of sight" transmission characteristics, an aircraft operating at a higher altitude can receive the signal of any given VOR at a greater distance. This could potentially cause problems if an aircraft was picking up transmissions from two different VORs sharing the same frequency, something more likely to happen while flying at altitude. Terminal VORs (TVORs) therefore are lower powered and frequency protected such that they are usable up to a distance of 25 nautical miles up to an altitude of 12,000' AGL. TVORs are not used for enroute navigation but are reserved for local navigation and instrument approaches.

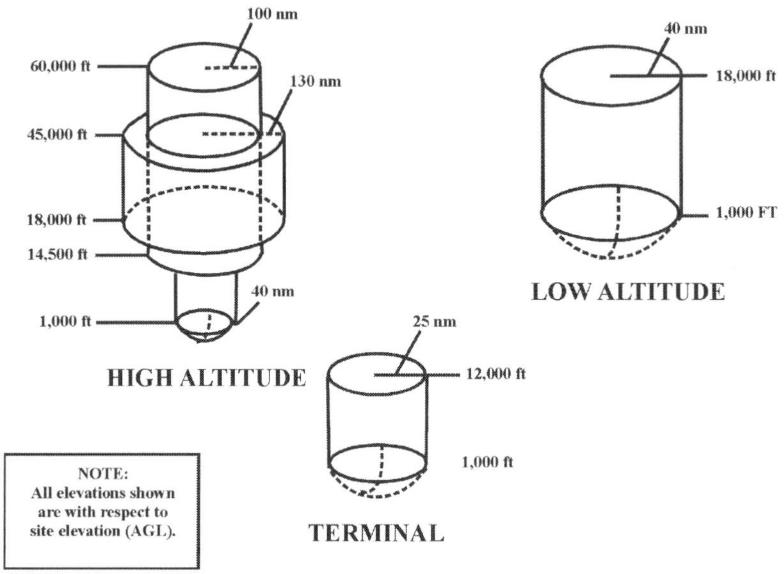

Figure 2-7. Terminal, low, and high altitude VOR service volumes

Every low-altitude VOR (L) is assigned an operating frequency that guarantee interference-free reception to aircraft operating up to 40 nautical miles away from the VOR at or below 18,000 feet. Low-altitude VORs cannot safely be used by aircraft operating above 18,000 feet or further than 40 miles away, as there is no guarantee that another VOR operating on the same frequency may not cause interference. High-altitude VORs (H) are used by aircraft operating between 18,000 and 60,000 feet and are guaranteed interference free reception at ranges up to 200 nautical miles. All these limitations imposed upon VORs are known as service volumes and are depicted in Figure 2–7.

Aircraft Positioning Methods

The VOR provides bearing information but cannot determine the aircrafts distance from the station. There is only one way for a pilot using VORs to accurately determine an aircraft's position: by cross referencing bearing information from two different VORs. By tuning in two different VORs, the pilot can plot a line of position (LOP) from each VOR. These two lines of position (or radials) can be drawn on a navigation chart, with the aircraft being located at the intersection of the two radials. This is a somewhat cumbersome and inaccurate process, and most pilots would prefer a system that could determine how many miles they are from any particular VOR. That system was invented and installed and is known as distance measuring equipment (DME).

DME Position Determination

If a pilot wishes to determine an aircraft's location using just one station, a system called DME must be used. DMEs are collocated with many VOR ground stations. When the pilot tunes in the appropriate VOR frequency, the DME equipment on the aircraft automatically tunes in the corresponding DME system. The pilot can then determine which radial the aircraft is located using the VOR receiver and uses DME to determine the aircraft's distance from the VOR. (See figure 2-8).

Figure 2-8. DME receiver/indicator

DME uses elapsed time measurement as the basis for determining distance from the VOR to the aircraft. The DME system consists of an interrogator located onboard the aircraft and a transponder located at the ground station. At regularly spaced intervals, the interrogator transmits a coded pulse on a frequency of around 1,000 MHz. When the ground-based DME transponder receives this pulse, it triggers a coded reply transmitted on a different frequency. When the interrogator on the aircraft receives this pulse, the elapsed time is electronically calculated. The measured time (known as range-time) is the total elapsed time between the transmission of an interrogation from the aircraft and the receipt of the reply to that interrogation. The approximate range time for a signal to travel one nautical mile and return is 12.36 microseconds. The DME equipment on board the aircraft measures the elapsed time between interrogator transmission and reception of that signal, divides by 12.36 micro-seconds, which provides distance from the ground station. This determination is known as the line of sight or slant range distance. Slant range is the actual distance between the aircraft and the ground-based DME transponder. (See figure 2-9).

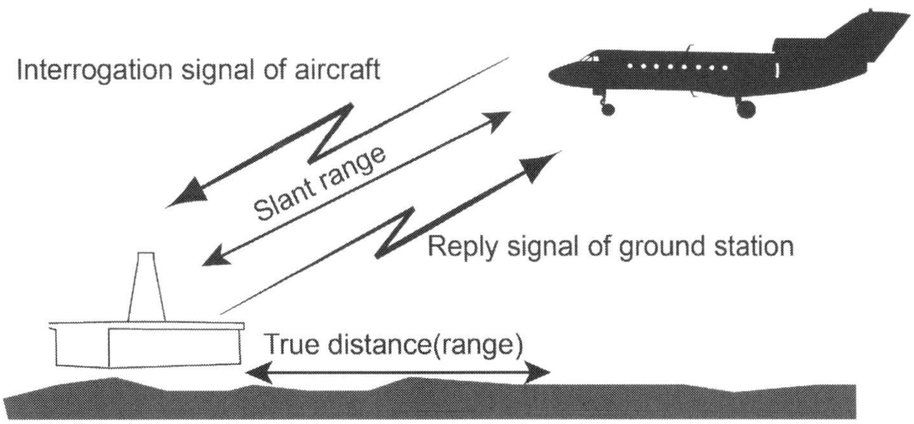

Figure 2-9. DME slant range distance

As the aircraft's altitude increases, the difference between slant range and ground distance increases. For instance, if an aircraft is 5.0 ground miles from the DME station, at an altitude of 6,000 feet, the DME indicator on board the aircraft will indicate approximately 5.1 nautical miles from the station. But if the aircraft is directly over the DME station, at an altitude of 30,000 feet, the DME indicator will also indicate about 5.1 nautical miles. The difference between slant range and ground distance is most pronounced when aircraft are operating at high altitudes fairly close to the DME ground station. This difference has been taken into consideration by the FAA when determining holding-pattern sizes, intersection locations, and airway positioning.

Tactical Air Navigation (TACAN)

The VOR-DME system has deficiencies that make it unusable for certain military operations. After an extensive evaluation of the civilian VOR-DME system, the Department of Defense chose to develop an alternative navigation system known as tactical air navigation (TACAN). TACAN is a polar coordinate–based navigation system that provides both bearing and distance information to the pilot using a single transmitter located on the ground or on a ship. TACAN does not use a passive transmitter on the ground like the VOR but instead operates in much the same way as the DME system.

During operation, the TACAN equipment on the aircraft (the interrogator) transmits a coded signal to the TACAN station on the ground (the transponder). On receipt of the interrogator signal, the transponder transmits a properly coded reply. The interrogator on board the aircraft measures the elapsed time and calculates the distance between the aircraft and the TACAN transmitter. (This is done in the same manner as with civilian DME equipment.) The interrogator on board the aircraft also decodes the signal and determines the aircraft's bearing from the TACAN ground station. The airborne equipment can then display both bearing and distance information to the pilot, using a display system similar to civilian VOR-DME indicators. TACAN eventually became the military navigation standard and is still in use today.

VORTAC

Instead of installing nationwide duplicate navigation systems that essentially provide the same capabilities, the FAA and the Department of Defense developed a multi-mode, combination VOR and TACAN based navigation system designed such that civilian aircraft use TACAN to provide distance information while still using VOR for azimuth information. Military aircraft are equipped solely with and only use the TACAN part of the equipment. This combined navigation aid is known as VORTAC and has become the world-wide navigation standard.

In recent years, as various newer navigation systems have come online, the FAA is decommissioning unneeded VOR/VORTAC stations. A skeleton system of VORTACs will remain in operation for the foreseeable future, however, as a potential backup/supplemental navigation system.

Area Navigation

When navigating airways using VORTAC, pilots are still required to fly from VORTAC to VORTAC until they reach the destination airport. Because of airport locations and VORTAC placement restrictions, it is seldom possible to navigate in a straight line from the departure to the destination airport. This forces pilots to fly a longer distance than necessary. It also creates congestion in the air traffic control system, since every aircraft is forced to navigate along a limited number of airways. In an attempt to alleviate this congestion, a number of systems have been developed to permit pilots to bypass the airway system and navigate directly to the destination airport. These various systems are collectively referred to as area navigation or RNAV.

The course-line computer (CLC) was developed to permit pilots to use existing VORTAC stations to fly directly from one airport to another. Using azimuth-distance navigation principles, the course-line computer determines the aircraft's current position using a standard VORTAC or VOR-DME station. The CLC on the aircraft then mathematically calculates the bearing and distance to any desired location (known as a waypoint) and produces navigation instructions that lead the pilot to that point. The distance and bearing to the waypoint are normally displayed to the pilot using the VOR course display and the DME distance indicator. The course-line computer was one of the most common area-navigation systems in use and was typically called RNAV. It has now been generally replaced by satellite-based RNAV systems.

Global Navigation Satellite System

The Global Navigation Satellite System (GNSS) is the accepted term for navigation systems that provide ground-based users with global navigation via space-based satellite systems. GNSS transmitters are typically located on low earth orbit satellites permitting users with fairly small, inexpensive receivers to determine their location in three dimensions (latitude, longitude, and altitude). As long as the receivers are within the sight line of a number of satellites, the users' receiver can determine their location within a few meters or even feet.

There are a number of competing GNSS systems, but all use the same general principles. Many GNSS receivers are designed to accept inputs from more than one system. The Global Positioning System (GPS), is the United States' GNSS system and is operated by the U.S. Air Force. Russia has its own GNSS system, known as GLONASS. The European Union is also developing their own system, called Galileo, as are China (BeiDou), Japan (Quasi-Zenith) and India (NAVIC). Due to its accuracy and worldwide availability, GNSS has been designated by ICAO as the current as well as future aviation navigation system.

Global Positioning System

GPS is a space-based positioning, velocity, and time system composed of a minimum of twenty four satellites in six orbital planes. The satellites operate in circular orbits arranged so that at any one time users worldwide are able to view a minimum of five satellites (see Figure 2–9). GPS operations are based on the concept of ranging and triangulation from a group of satellites in space that act as precise reference points.

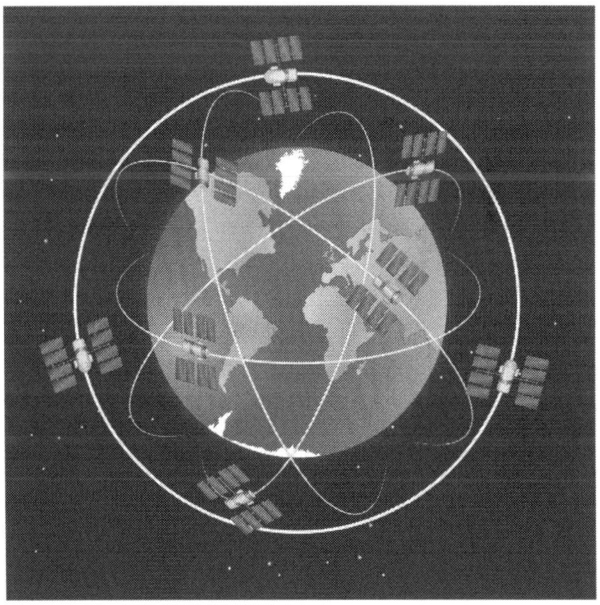

Figure 2-10. GPS satellite constellation

A GPS receiver determines location by comparing the travel time of radio signals from multiple satellites. Each satellite transmits a specific code, called a course/acquisition (CA) code that contains information on the satellite's position, an extremely precise GPS system time, clock error, and the accuracy of the transmitted data. The GPS receiver on the aircraft matches each satellite's CA code with an identical code contained in the receiver's database. By comparing the reception time from each satellite, the receiver can calculate its position relative to each satellite. The aircraft GPS receiver then mathematically determines its position by triangulation. Using stored information, the GPS receiver can also compute navigational information such as distance and bearing to a point, ground speed, estimated time enroute, estimated time of arrival, and winds aloft. It does this by using the aircraft's known latitude/longitude, measuring relative movement, and referencing these to a database built into the receiver. (See figure 2-11).

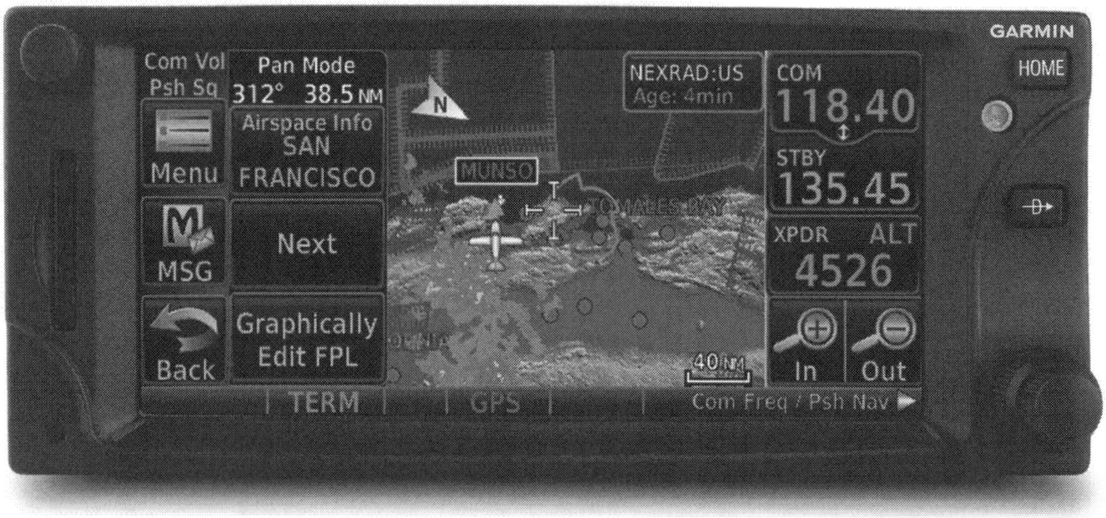

Figure 2-11. GPS receiver/indicator

GNSS Augmentation

GNSS signals provide accuracy for enroute and two-dimensional navigation, but they do not provide acceptable vertical or lateral guidance for all weather landings. The standard GNSS signal needs to be augmented for aircraft to have this capability. There are three general augmentation schemes approved for use; the Satellite Based Augmentation System (SBAS), Ground Based Augmentation System (GBAS), and the Aircraft Based Augmentation System (ABAS).

Augmentation provides more accurate lateral guidance during the approach and departure phases of flight and can be used in some enroute environments as well. Augmentation can also provide approach with vertical guidance (APV), which offers pilots a positive and stabilized vertical guidance flight path for approach procedures where no current guidance exists.

Satellite Based Augmentation Systems

The SBAS system comprises a network of ground reference stations that collect satellite signals and send them to one or more ground processing centers. These centers compare the overall signal inaccuracy from each station, compute a differential correction for a general geographic area, then send a correction signal to one or more geostationary satellites that transmit the augmentation messages to aircraft using GNSS.

There are multiple SBAS systems being developed and/or in operation. Most operate regionally and encompass each country's airspace. One SBAS currently operational is the U.S. wide area augmentation system (WAAS).

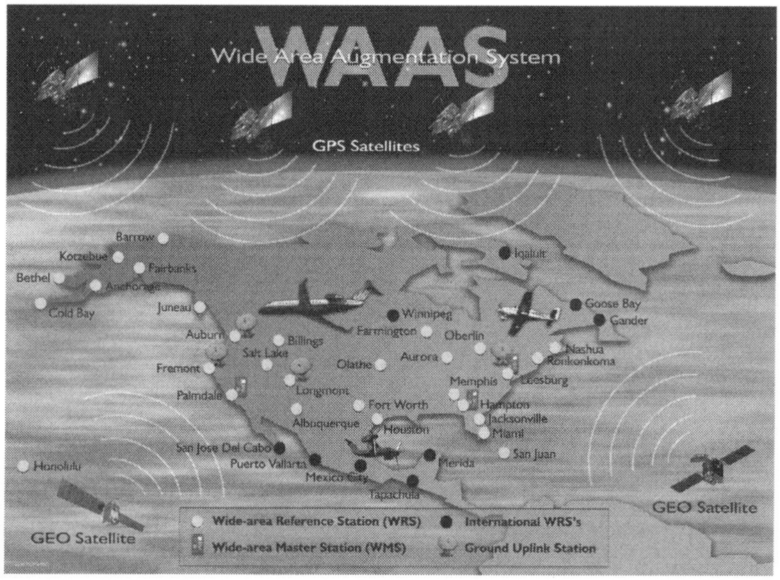

Figure 2-12. Wide Area Augmentation System, satellite based GPS augmentation system

WAAS uses a network of precisely located ground reference stations to monitor GPS satellite signals. These stations are located throughout the United States (primarily co-located with ARTCCs) as well as in Canada and Mexico. These ground reference stations collect and process error information and send it to the three WAAS master stations located in Georgia, California and Virginia. The master stations more or less "average" the error and develop a correction message sent to users via geostationary satellites located above the United States. Using WAAS, GPS signal accuracy is improved from about plus or minus 20 meters to approximately 2 meters both horizontally and vertically (see Figure 2–12).

Ground-Based Augmentation System (GBAS)

Aircraft using GBAS receive augmentation information directly from a local ground-based transmitter. GBAS is similar to SBAS with the exception that the system error is measured, corrected and transmitted in only one local geographic area (about 30 nm radius), thereby making the augmentation differential calculation extremely accurate. The augmentation message is only sent to aircraft in the local area, usually by some form of terrestrial radio communication. (See figure 2-13)

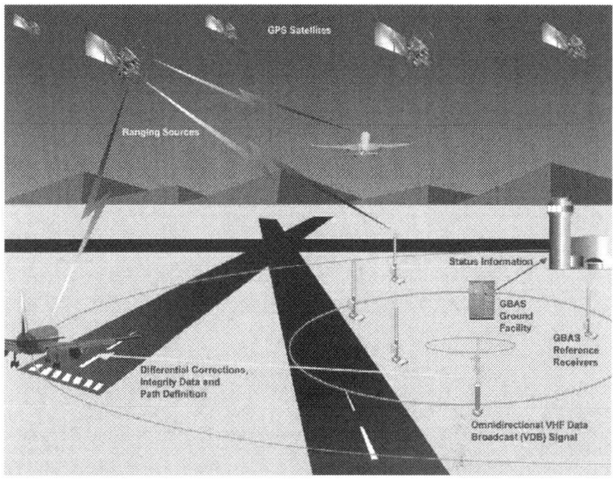

Figure 2-13. GBAS-ground based GPS system

The U.S. operated GBAS system used to (and is occasionally still) referenced as the Local Area Augmentation System, or LASS. The official name for the LASS system is now GBAS. If a LASS system is an integral part of a precision landing system, it is known as the GBAS Landing System, or GLS. GBAS and the associated GLS procedures are still in the development stage, but GLS instrument approaches have been developed and are being utilized ant Newark and Houston international airports as well as at about 20 additional airports world-wide.

Aircraft Based Augmentation System

Aircraft based augmentation systems (or ABAS) are still under development but are designed around the premise of using independent navigation systems onboard each aircraft to augment the GNSS satellite signal. ABAS is still in the early development and regulatory stage, but it appears likely to utilize additional positional information derived from onboard navigation sensors to augment the performance of the GNSS system. A couple of possible ABAS solutions includes Receiver Autonomous Integrity Monitoring (RAIM), which uses additional and redundant GPS signals to ensure the integrity of the aircraft position solution. The use of e-LORAN, Inertial Navigation Systems (INS) and even automated celestial navigation systems are under consideration. All of these systems will eventually be brought together under one umbrella program known as Required Navigation Performance or RNP.

Required Navigation Performance

Instead of mandating sensor specific (such as VORTAC or GPS) navigation standards and developing routes and approaches for each specific type of sensor, the RNP concept simply provides specifications based on demonstrated levels of navigation performance and capabilities. The means of meeting these (various) levels of performance are left up to the manufacturer and the aircraft operator. In general, the more equipment installed, the better the navigation capability of the aircraft. The RNP concept does not mandate the use of any specific navigation system or equipment, but instead simply specifies an accuracy/performance level that must be maintained to use certain airways or approach procedures.

RNP navigation performance accuracy is quantified as two values: a distance in nautical miles (known as the RNP type) and a probability level (usually 95%). For example, an airplane would be certified to operate on an RNP-2 airway if the performance of the installed navigation system would result in the airplane being within 2 nautical miles of its indicated position at least 95% of the time.

The RNP capability of an aircraft varies depending upon the equipment installed on the aircraft as well as the navigation infrastructure. When the system is fully mature, RNP authorized aircraft will likely be equipped with multi-mode receivers (MMR) that select the most accurate system available to the receiver then displays appropriate navigation information to the pilot. The pilots can then use any procedure for which the aircraft's navigation systems qualify. For example, an aircraft equipped to an RNP-2 standard could fly along any airway (or approach procedure) designated as RNP 2.0. Aircraft could be equipped with various sensors that could enable accuracy as good as RNP-.1. During flight, the onboard MMR will automatically select from GNSS, VOR, TACAN, ILS, or DME navigation signals to provide the pilot with the most accurate solution set. (See figure 2-14)

Figure 2-14. Garmin 1000 integrated navigation and display system

Different airspace, routes, or procedures will have specific minimum RNP requirements. ICAO has already defined standard minimum RNP values for four navigation phases of flight: oceanic, enroute, terminal, and approach. The required RNP value is expressed as a distance in nautical miles from the intended centerline of a procedure, route, or path.

TABLE 2-1 RNP NAVIGATION STANDARDS

RNP Level	RNP Minimum Accuracy	Phase of Flight
.1 to 1	0.1 nm-1.0 nm	Approach and landing
.3	0.3 nm	Approach segments
1	1.0 nm	Terminal area and some enroute
2	2.0 nm	Enroute

Instrument Approach Procedures

The navigation systems previously discussed are primarily utilized for enroute navigation. If, upon arrival at the destination airport, the pilot can see the airport and safely approach and land, the pilot may use what is called a visual approach. In general, a visual approach can be conducted if the visibility is 3 miles or greater and the pilot can either see the airport or can follow another aircraft to the airport. During a visual approach, the pilot accepts the responsibility for navigating to the airport and avoiding any obstacles within the local area. When visual approaches are being conducted, air traffic controllers are still responsible for separating aircraft that are using them from aircraft operating on IFR flight plans but the navigation is left to the pilot.

If the weather conditions at the destination airport are such that the pilot is unable to, or chooses not to, conduct a visual approach, he or she must conduct an instrument approach procedure (IAP). During the conduct of an instrument approach, the pilot must follow a specified procedure that provides course guidance and obstacle clearance. This procedure guides the pilot to the destination airport where he or she can then make a safe landing.

Segments of an Instrument Approach Procedure

An instrument approach procedure essentially consists of four components: the initial approach, intermediate approach, final approach, and missed approach segments. (See figure 2-15)

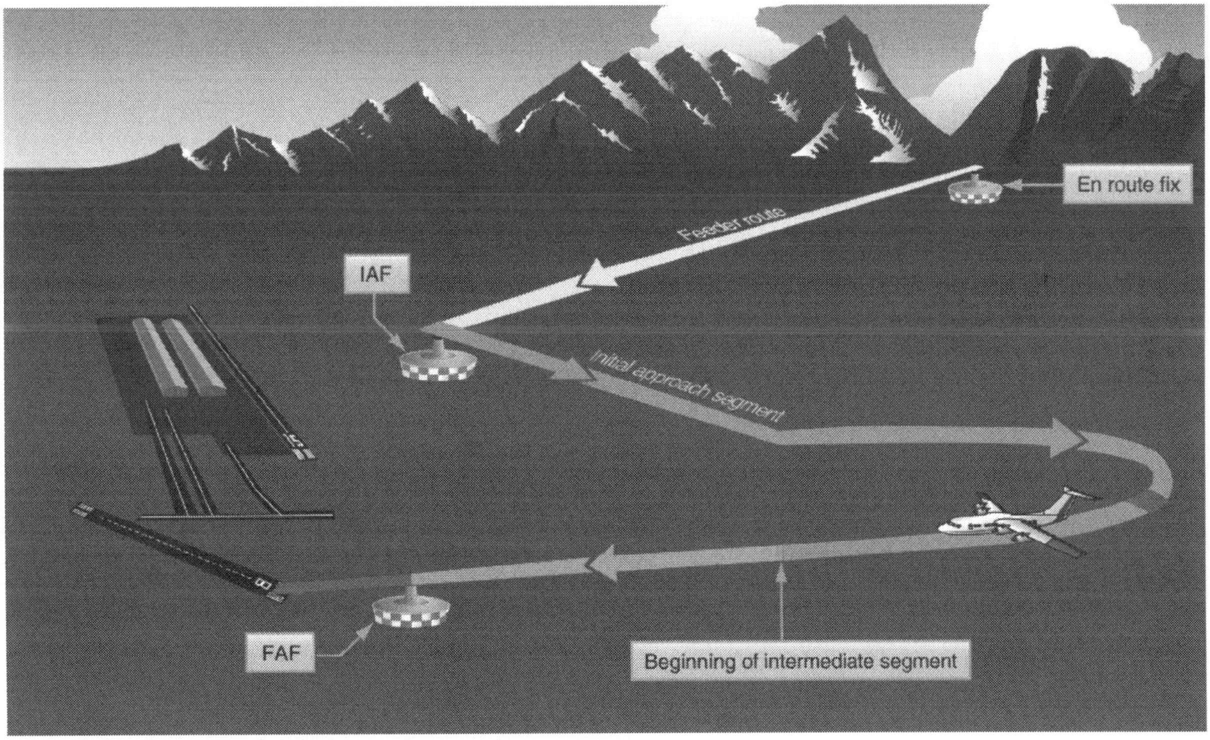

Figure 2-15. Segments of an instrument approach

Initial Approach Segment

The initial approach segment is designed to transition the aircraft from the enroute airway structure to the intermediate approach segment. The initial approach segment begins at one of the initial approach fixes (IAFs) located along the federal airways. This segment is usually defined as a heading or a radial to fly from the IAF to the beginning of the intermediate approach segment. Each initial approach segment typically specifies the minimum allowable altitude that may be flown along that route. There is usually one initial approach segment for every airway that pilots might be using as they approach the airport. The initial approach segment terminates when it joins the intermediate approach segment.

Intermediate Approach Segment

The intermediate approach segment is designed to permit the pilot to descend to an intermediate altitude while aligning the aircraft to make an easy transition to the final approach segment. The intermediate approach segment terminates at the final approach fix (FAF), which is designated on the approach chart with a maltese cross for nonprecision approaches and a lightning bolt for precision approaches. There is usually only one intermediate approach segment for every approach. It is not ordinarily identified as such on an approach chart.

Final Approach Segment

The final approach segment is used to navigate the aircraft to the runway and properly position it to permit a safe (visual) landing. This segment begins at the final approach fix and ends at the runway or at the missed approach point (MAP). The final approach segment guides the aircraft to the desired runway using a navigation aid located either at the airport or nearby. The approach procedure can be one of two general types: precision or nonprecision. A nonprecision approach provides only lateral guidance to the pilot. A precision approach provides the pilot with both lateral and vertical course guidance to the approach end of the runway.

Nonprecision Approach

During a nonprecision approach, after crossing the final approach fix the pilot descends to a predetermined minimum descent altitude (MDA). The pilot levels off and maintains this altitude while tracking along the final approach segment toward the missed approach point. If the runway or runway environment is sighted prior to reaching the MAP and the pilot feels that a safe landing can be made, he or she is legally authorized to continue the approach and land. If the runway is not in sight prior to reaching the MAP, or if a safe landing cannot be accomplished, the pilot must transition to the missed approach segment, which usually leads back to an initial approach fix. This is called a missed approach procedure. Typically, a non-precision approach procedure will enable the aircraft to descend to around 300' above the ground, and usually requires a landing visibility of about one statute mile.

Precision Approach

During a precision approach, the pilot is provided positive vertical navigation guidance while flying the final approach segment. A precision approach provides either a transmitted electronic or a computer calculated descent path for the pilot. This is known as a glide path. When the designated altitude (if measured above sea level it is known as the decision altitude or DA, if measured above the ground it is decision height or DH) has been reached, the pilot must determine whether a safe landing can be made. If it is safe to land, the pilot continues down the glidepath and lands. However, if the pilot determines that it is not safe to continue, a transition to the missed approach segment must be made, and the missed approach procedure must be conducted.

Because a precision approach aid provides both lateral and vertical guidance, the pilot is usually authorized to descend to a lower altitude before making a decision about landing. Typically, a precision approach procedure enables the aircraft to descend to around 200' above the ground, and usually requires a landing visibility of about one-half statute mile, although there are some special (expensive) systems that permit lower landing minimums. Precision approaches are much more valuable to the pilot during periods of marginal weather. Since precision approach aids are usually more expensive to purchase, install, and operate than nonprecision aids, they are normally reserved for use at airports that experience a significant amount of marginal weather conditions.

Approach Navigation Aid Classifications

Although GNSS is becoming the dominant form of aviation navigation, there is still a need for non-GNSS approach navigation solutions. As a significant number of aircraft (primarily smaller private aircraft as well as some military models) do not have extensive GNSS capabilities, non-satellite-based approach procedures are still needed in case of a GPS system malfunction or failure.

Many enroute navigation aids can be used as nonprecision approach aids if their transmitted signal is of a high enough quality and can be safely used during the entire instrument approach procedure. FAA flight-check aircraft routinely check the quality of these navigation aids to determine their suitability as approach navigation aids. An enroute navigation aid used for an instrument approach is classified as a nonprecision aid since no vertical guidance is provided to the pilot. The following enroute navigation aids have been certified by the FAA for use as nonprecision approach aids:

VOR	RNAV	VOR-DME
GPS	VORTAC	TACAN

The FAA has developed an entire series of radio navigation devices to serve solely as instrument approach aids. Most of these qualify as precision approach aids since they provide vertical guidance to the runway. The instrument landing system (ILS) is the most common.

Instrument Landing System

The ILS is designed to provide the pilot with an approach path perfectly aligned with the runway centerline as well as a glidepath leading down to the runway. The ILS system is equipped with three different types of transmitters: the localizer, the glide slope, and two or three marker beacons. (See figure 2-16)

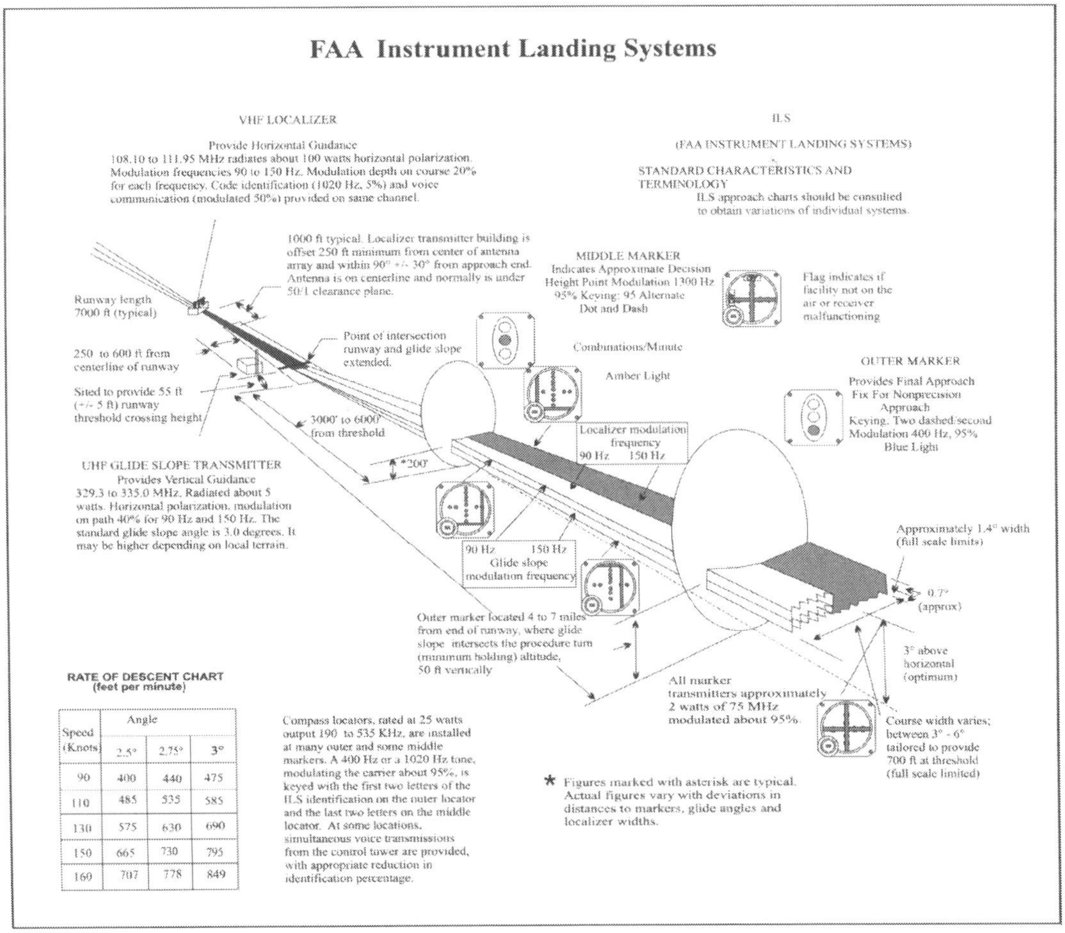

Figure 2-16. Instrument Landing System

RATE OF DESCENT CHART (feet per minute)

Speed (Knots)	Angle		
	2.5°	2.75°	3°
90	400	440	475
110	485	535	585
130	575	630	690
150	665	730	795
160	707	778	849

Localizer

The localizer consists of a transmitter building, localizer antenna, and monitoring equipment. The localizer operates within the VHF band between 108.10 and 111.95 MHz and provides the pilot with lateral course guidance. Left-right guidance is provided to the pilot by a vertical needle in a manner similar to that used by the VOR indicator. In fact, in many aircraft installations, the VOR indicator is also used for the ILS.

Glide Slope

The glide slope radiates a signal pattern vertically to provide an electronic glide path to be flown by the pilot. The glide slope system provides both above and below glide path indications to the pilot, using a horizontal needle on the ILS indicator. The glide slope operates in the UHF band between 329 and 335 MHz and is paired to the localizer frequency. When a pilot selects the proper localizer frequency, the glide slope frequency is also selected by the ILS receiver.

Marker Beacons

Marker beacons are located at known distances along the final approach course of the ILS and provide position information to pilots conducting the approach. Marker beacons transmit a cone-shaped signal on a frequency of 75 MHz, uniquely coded to identify each type of beacon. Outer marker (OM) beacons are usually located on the ground about 5 miles from the approach end of the runway. When an aircraft flies over an outer marker, a blue light flashes and a 400 Hz series of continuous dashes is emitted from the marker beacon receiver. The middle marker (MM) is about 3,000 feet (or half a mile) from the approach end of the runway and causes an amber light to flash and a series of 1,300 Hz dots and dashes to be heard in the cockpit. The middle marker is usually located such that an aircraft properly positioned on the glide slope will overfly it at approximately 200 feet. This is the normal decision height for a Category I ILS approach.

ILS Categories

ILS systems are currently classified into one of three categories, each category being defined in terms of minimum visibility and decision height altitudes. Minimum visibility is measured in fractions of a mile when measured by human observers or in hundreds of feet when measured by a runway visual range. The standard ILS is a Category I, which provides accurate guidance information in visibilities as low as 1/2 mile and ceilings as low as 200 feet. These minima are representative of a standard ILS installation. (See figure 2-17)

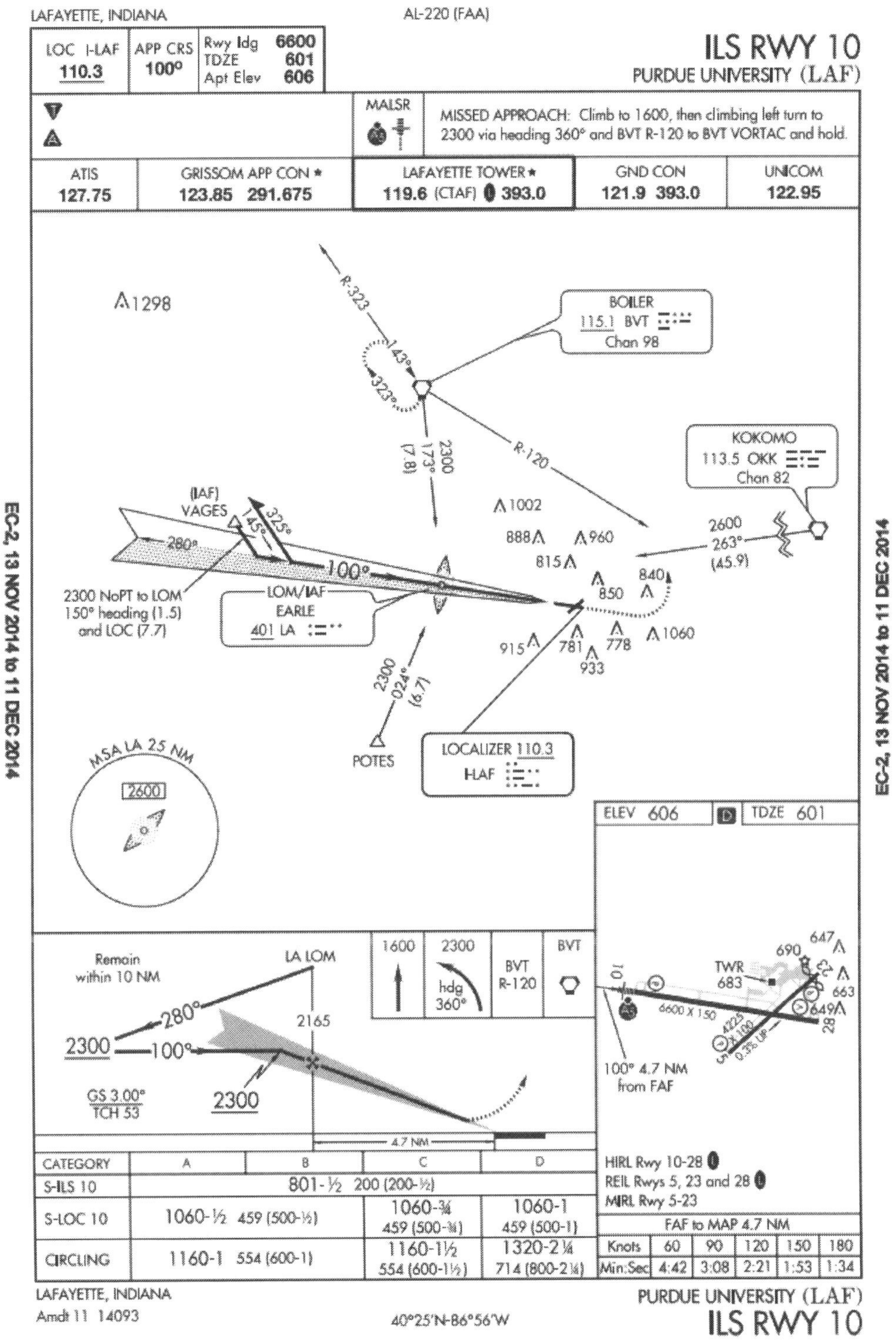

CATEGORY	A	B	C	D
S-ILS 10	801-½ 200 (200-½)			
S-LOC 10	1060-½ 459 (500-½)		1060-¾ 459 (500-¾)	1060-1 459 (500-1)
CIRCLING	1160-1 554 (600-1)		1160-1½ 554 (600-1½)	1320-2¼ 714 (800-2¼)

FAF to MAP 4.7 NM

Knots	60	90	120	150	180
Min:Sec	4:42	3:08	2:21	1:53	1:34

Figure 2-17. ILS approach chart for Purdue University airport in Lafayette, Indiana

With a slight change in the ground equipment, an ILS installation may be certified as Category II, which permits a properly rated pilot to land in visibilities as low as 1,200 feet or ceilings as low as 100 feet. The additional equipment required for a Category II installation includes more stringent localizer and glide slope monitoring equipment, an inner marker, and additional approach lighting. Pilots and aircraft must be certified to use a Category II ILS and its associated minima. Those pilots not certified to Category II minima may still use a Category II ILS using Category I minima. (See figure 2-18)

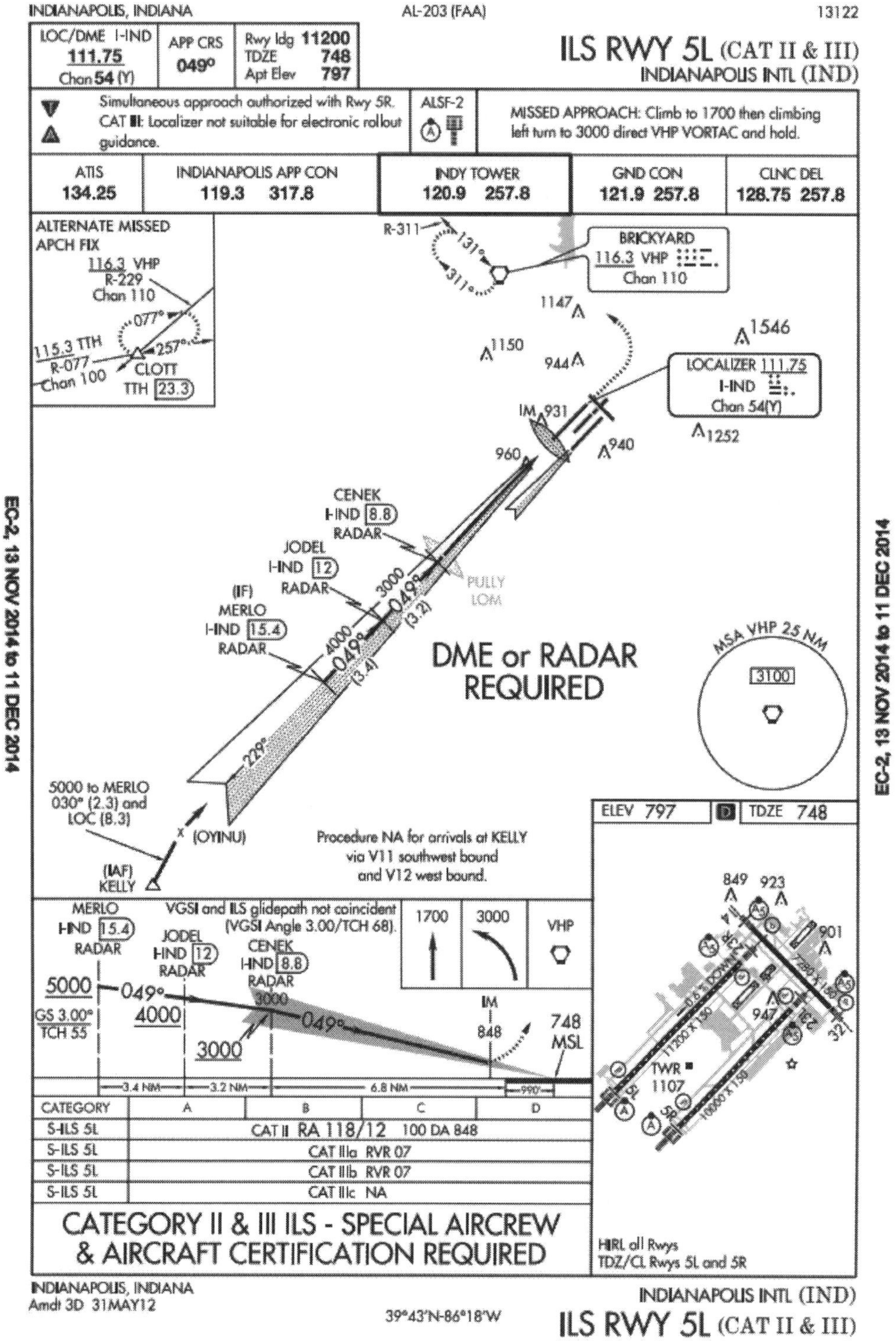

Figure 2-18. Category II and III ILS approach chart for Indianapolis, Indiana

At those airports that qualify, a Category III ILS may be installed. A Category III ILS is much more expensive to install and operate as it requires a completely redesigned localizer and glide slope equipment and more extensive approach and runway lighting. Category III ILS approaches can provide three different minima and are classified as category IIIa, IIIb, or IIIc, each with progressively lower minima. Category IIIc approaches may be conducted when either or both the ceiling or visibility is zero. Aircraft conducting Category III approaches must be equipped with autoland devices that automatically land the aircraft. Category III installations are rarely justified for use in this country. Few airports need this type of approach and few aircraft are equipped to utilize them.

Table 2-2 ILC Approach categories

ILS Category	Minimum Visibility	Decision Altitude (height above the ground)
I	1,800’ or 1/2 mile	200’
II	1,00’	100’
IIIa *	700’	none
IIIb *	150’	none
IIIc *	none	none

**- Category III approaches require special onboard equipment. As of 2014, no category IIIc approaches have been approved in the U.S.*

GPS-Based Instrument Approaches

GNSS is becoming the international standard for instrument approaches. As part of the transition process from ground-based navigation aids, the FAA started with a GPS overlay approach program. This process permits pilots to use GPS to fly certain nonprecision instrument approach procedures. These procedures are identified by name, and the phrase “or GPS” is then added to the title. For example, “VOR/DME or GPS RWY 27L”. As the development of stand-alone GPS approaches has progressed, many of the original overlay approaches have been replaced with stand-alone procedures specifically designed for use by GPS systems. The title of these procedures have only the GPS navigation system in the title. For example, “GPS RWY 24”.

Waypoints/Computer Navigation Fix

GPS receivers are designed to guide pilots to GPS defined intersections known as waypoints. To differentiate these intersections from those used by the VOR system, a new system of identifiers has been created. Under this naming system, any point used for the purpose of defining the navigation track of an aircraft is called a Computer Navigation Fix or CNF. The FAA has begun to assign five-letter names to CNFs and to chart these on various aeronautical charts. CNFs cannot be used for any air traffic control application, such as holding or re-routing of aircraft, but are names assigned to waypoints that are included in an aircraft’s internal navigational database.

In most cases, CNF names are unique, with the exception of some waypoints associated with the runway itself. For example, some runway threshold waypoints, which are generally the beginning of the missed approach segment, may be assigned a unique, five-letter identifier, but they may also be coded with a runway number such as “RW36L”.

Approach and Landing Procedures

As stated previously, if an IFR pilot can navigate from the airway to the airport by sight, they may conduct a visual approach. But if that is not possible, the pilot must conduct an instrument approach procedure (IAP). In the past, these various procedures were identified by the type of navaid upon which it was based (for example, VOR, ILS, NDB, TACAN, GPS, etc.). In reality, there are only three basic categories of instrument approaches currently in use in the national airspace system.

TABLE 2-3 TYPES OF INSTRUMENT APPROACHES

Approach	Guidance provided	Minimum altitude (height above the ground)	Minimum visibility	Navigation Aids
Nonprecision approaches	Lateral only	300'-500'	1 statute mile	VORTAC GPS ILS (localizer only)
Category I precision approach	Lateral and Vertical	200'-300'	½ to 1 statute mile	ILS GPS (WAAS augmented)
Category II/III approach	Lateral and Vertical	Less than 200'	Less than ½ mile	ILS GPS (GBAS augmented)

These categories will eventually be replaced by three different categories of GNSS based procedures. These categories will similarly be defined by their navigational accuracy.

TABLE 2-4 INSTRUMENT APPROACH CAPABILITIES AND ACCURACIES

Approach	Guidance provided	Course deviation indications	Type of Approach	Augmentation needed	Similar to existing
Lateral Navigation (LNAV)	Lateral guidance only. No vertical guidance is provided.	Course deviation indications are linear during the entire approach.	Non-Precision. Minimum altitudes are expressed as MDAs.	WAAS required	Similar to current non-precision approaches. There are relatively few of these approaches.
Lateral Navigation/Vertical Navigation (LNAV/VNAV)	Lateral and vertical guidance provided. LNAV/VNAV certified aircraft are permitted to use GPS or Baro-VNAV to calculate vertical	Course deviation indications are linear during the entire approach.	Precision. Minimum altitudes are expressed as DAs.	Lateral Navigation/Vertical Navigation (LNAV/VNAV)	Lateral and vertical guidance provided. LNAV/VNAV certified aircraft are permitted to use GPS or Baro-VNAV to calculate vertical

HALF PRICE BOOKS®

Half Price Books
1835 Forms Drive
Carrollton, TX 75006
OFS OrderID 39482158

Thank you for your order, ABE TP Orders!

Thank you for shopping with Half Price Books! Please contact Support@hpb.com. if you have any questions, comments or concerns about your order (738046715)

Visit our stores to sell your books, music, movies games for cash.

SKU	ISBN/UPC	Title & Author/Artist	Shelf ID	Qty	OrderSKU
S471471122	9780996245265	A Career in Air Traffic Control Paperback Michael S. Nolan,Sharon L. LaRue	66--01--3	1	

SHIPPED STANDARD TO:
ABE TP Orders
800 Avondale Ave
Columbus OH 43212
help@textbookrush.com

ORDER# **738046715**
Abebooks

Approach	Guidance provided	Course deviation indications	Type of Approach	Augmentation needed	Similar to existing
	guidance.				guidance.
Localizer Performance (LP)	Lateral guidance only	Course deviation indications are similar to the ILS, getting more sensitive as the aircraft approaches the runway.	Non-Precision. Minimum altitudes are expressed as MDAs.	WAAS required	Similar to an ILS approach without the glideslope.
Localizer Performance with Vertical guidance (LPV)	Lateral and vertical guidance	Course deviation indications are similar to the ILS, getting more sensitive as the aircraft approaches the runway.	Precision. Minimum altitudes are expressed as DAs.	WAAS required	Similar to an ILS approach.
GNSS Landing System (GLS)	Lateral and vertical guidance	Course deviation indications are similar to the ILS, getting more sensitive as the aircraft approaches the runway.	Precision. Minimum altitudes are expressed as Decision Altitudes.	GBAS required	Similar to ILS category II approaches

Sample GNSS based instrument approach charts are provided as Figure 2-19 and 2-20.

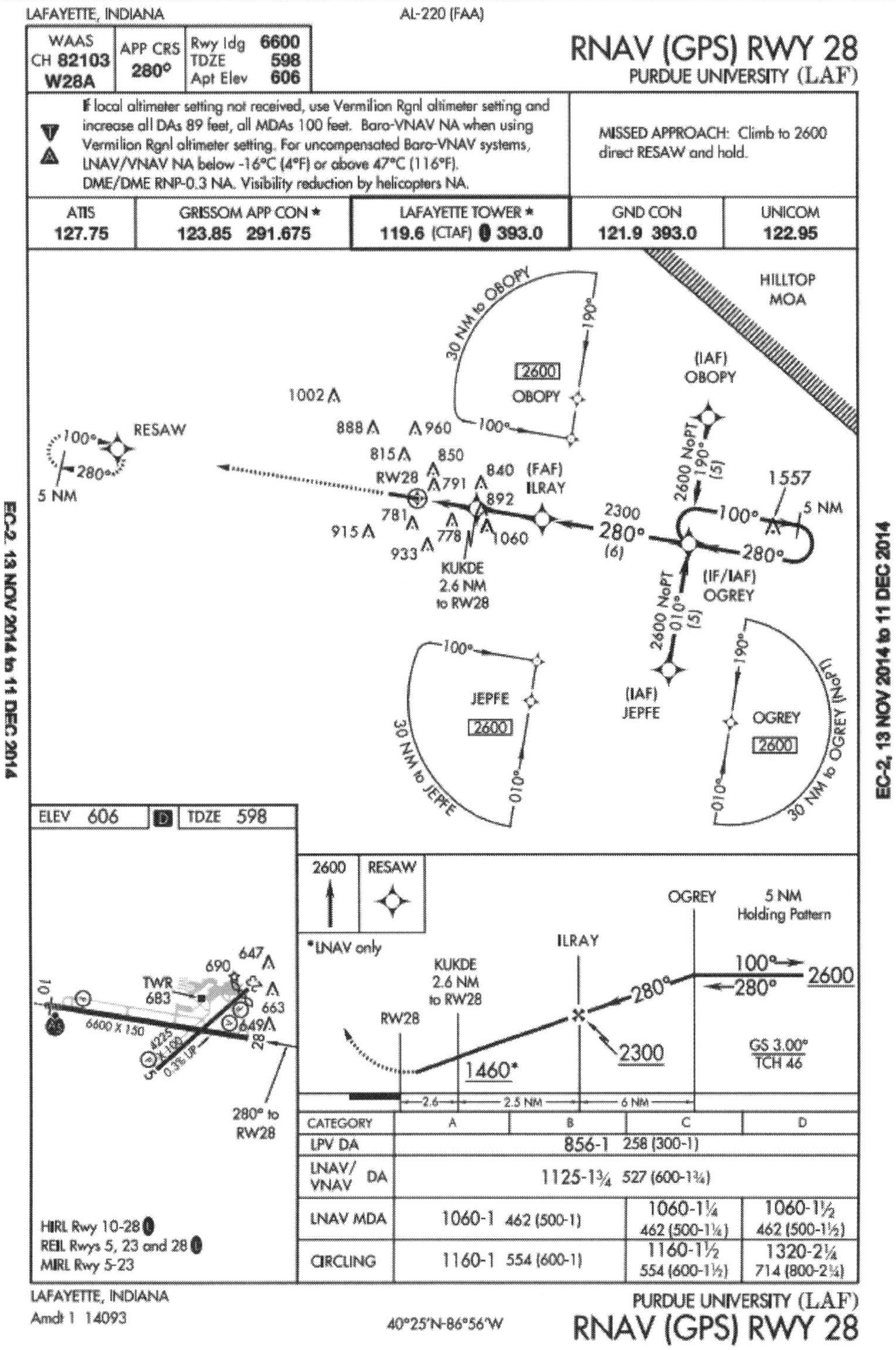

CATEGORY	A	B	C	D
LPV DA	856-1 258 (300-1)			
LNAV/VNAV DA	1125-1¾ 527 (600-1¾)			
LNAV MDA	1060-1 462 (500-1)		1060-1¼ 462 (500-1¼)	1060-1½ 462 (500-1½)
CIRCLING	1160-1 554 (600-1)		1160-1½ 554 (600-1½)	1320-2¼ 714 (800-2¼)

Figure 2-19. GPS approach for the Purdue University airport in Lafayette, Indiana

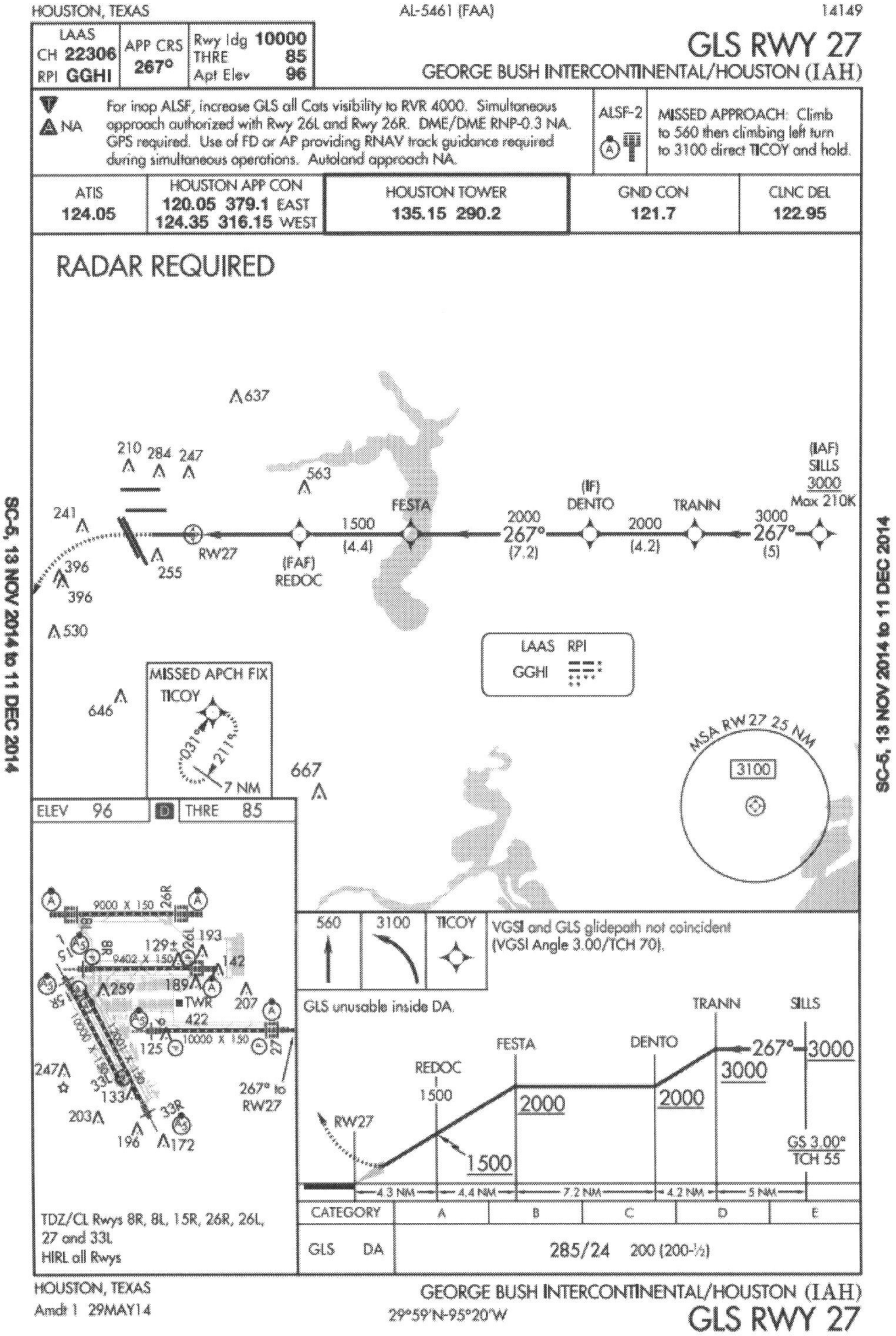

Figure 2-20. GLS approach for the Houston-George Bush Intercontinental Airport

Airport Markings and Signage

Aeronautical navigation is not restricted to inflight-electronic navigation systems. Navigation on the ground is just as important and a failure to navigate correctly at an airport can lead to disastrous consequences. Airports world-wide have adopted a comprehensive system of identification, lighting and marking of runways and taxiways to make the airports more efficient and safer to operate on.

Runway Layout and Numbering

Runways are referred to by a number between 1 and 36 corresponding to the direction of that runway in relation to magnetic north, rounded to the nearest 10 degrees. For example, a runway heading east (90°) would be referred to as runway 9. A runway headed west (270°) would be runway 27, one headed north would be runway 36 while one headed south would be runway 18.

Each end of the runway has a different number, 180 degrees opposed to the opposite side.
If there is more than one runway headed in the same direction each runway is identified by appending an appropriate name (left, right or center) to the runway number. For example, the left most runway headed north would be called runway 35 Left (L). The other northerly runway would be 35R. If there were three runways, the middle one would be called runway 35C (center). (See figure 2-21).

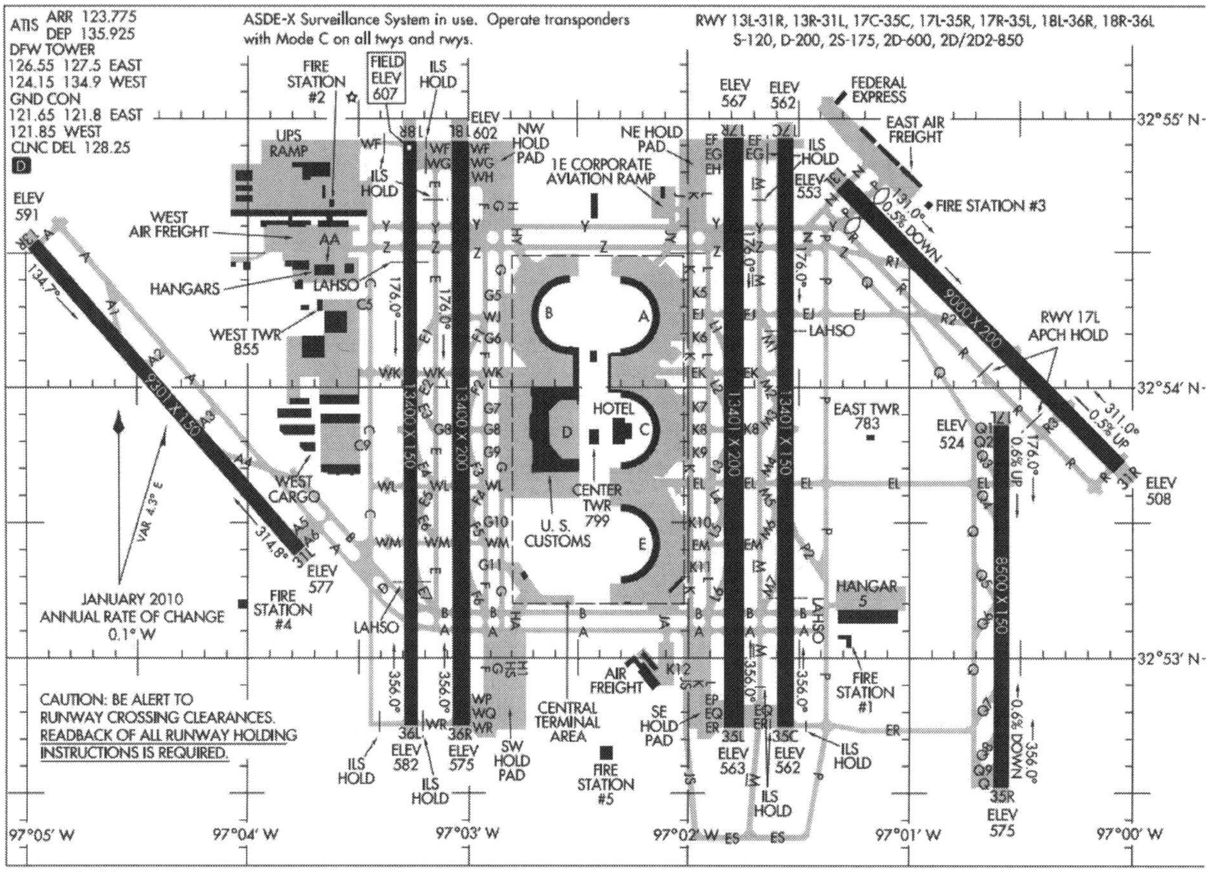

Figure 2-21. Dallas-Ft. Worth runway layout

Since there are only three possible variations using this type of runway classification, an airport with four or more parallel runways will need to have one set of runway identifiers shifted by 10 degrees. For example, at Dallas-Fort Worth, this results in runways 36L, 36R, 35L, 35C and 35R, even though all five runways are parallel. When referring to the runway verbally, each digit in the runway name is pronounced individually. In the United States the leading zero is dropped in all communications and runway markings. For example, a runway heading 70 degrees would be referred to as runway 7.

Pavement Markings

Runway pavement markings are almost always white, which is a visual cue to the pilot that the pavement is to be used for landing and departing. Most runways are outlined with a single solid white line. The runway number itself is painted on each end. A series of parallel vertical stripes are placed on the end (threshold) of each runway. The stripes are a standard width which helps the pilot estimate the width of the runway based on the number of vertical stripes.

The runway centerline is identified using a line of uniformly spaced stripes and gaps similar to that found on a highway. There are markings used to identify the touchdown zone for landing operations and are coded to provide distance information to the pilot in 500 feet increments. Touchdown zone markings consist of groups of one, two, and three rectangular bars symmetrically arranged in pairs about the runway centerline. There is also a set of large, white rectangles painted on the runway surface about 1000’ down the runway known as the fixed distance marker or runway aiming point. These large rectangular markings act as an aiming point for the pilots while landing. (See figure 2-22). Sometimes construction, maintenance, or other activities require the runway threshold to be temporarily relocated further down the runway. If the relocation is permanent, it is called a displaced threshold and marked accordingly.

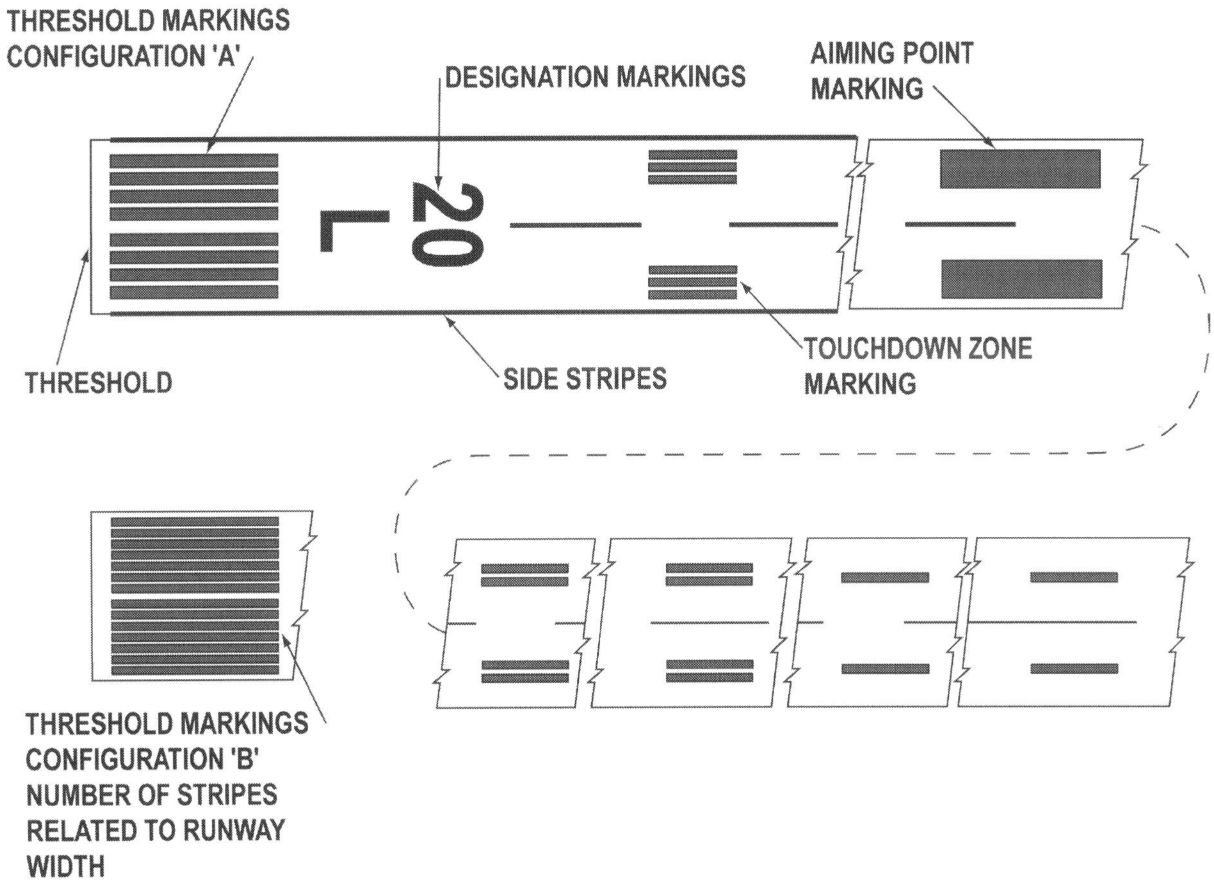

Figure 2-22. Typical runway markings

Taxiways

Taxiways are used by aircraft to travel to and from the runways. Taxiways are generally labeled using letters of the alphabet (alpha, bravo, charlie, etc.) and are marked with yellow lines on both sides of the taxiway as well as the centerline. Taxiways also have runway holding position markings wherever they intersect a runway. (See figure 2-23). Taxiways, holding areas, and parking aprons are sometimes provided with paved shoulders to prevent blast and water erosion. Although shoulders may have the appearance of full strength pavement they are not intended for use by aircraft and may be unable to support an aircraft. Usually the taxiway edge marking will define this area. Where conditions exist such as islands or taxiway curves that may cause confusion as to which side of the edge stripe is for use by aircraft, taxiway shoulder markings may be used to indicate the pavement is unusable. Taxiway shoulder markings are yellow.

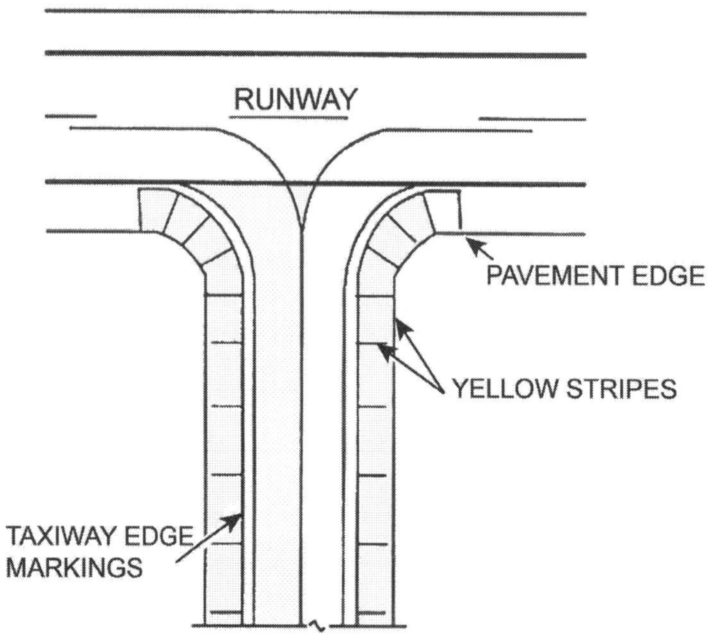

Figure 2-23. Typical taxiway markings

Runway and Approach Lighting

When night flying was first introduced, most airports consisted of an open area covered with either turf or cinders. Although visual navigation to the airport worked well during daylight hours, at night or in marginal weather conditions it was difficult for pilots to see objects on the ground and make an accurate determination of their aircraft's position. Sparsely populated areas of the country did not offer sufficient ground references to permit the pilot to determine the aircraft's location. If and when the pilot finally arrived at the destination airport, he or she might find it difficult to actually locate the runway in the dark and land the aircraft. The solution was to develop and install both airway and airport lighting.

The first airway lighting was instituted in the 1920s. At equal intervals along the airway, rotating beacon lights were installed that delineated the airway's center line. These rotating beacons were installed on steel towers and consisted of 1,000-watt electric lamps housed in a rotating drum assembly equipped with 36-inch-diameter lenses at each end. One lens was clear while the other lens was colored. These beacons were installed along the airway at 15-mile intervals. As the pilots flew along the airway, the beacons would appear as flashes of light visible from a distance of over 40 miles. To visually navigate along the airway, all the pilot needed to do was to fly from one beacon to the next.

Each rotating beacon was equipped with a colored lens that uniquely identified that particular beacon and enabled pilots to accurately determine their position. Each airport along the airway was also equipped with a rotating beacon having one clear and one green lens. These beacons were designed to help pilots determine the airport's exact location. Although a few airway beacons still exist in mountainous regions in the western United States, the green and white rotating beacon remains in use at civilian airports today. Other color combinations are used to differentiate other types of airports. The assigned colors for airport rotating beacons are as follows.

TABLE 2-5 ROTATING BEACON COLOR COMBINATIONS

Beacon Colors	Airport Type
White and green	Land airport
White and yellow	Water airport
Green, white, and white	Military airport
Green, yellow, and white	Heliport
Green, red, and white	Hospital heliport

Runway Lighting

In the early days of aviation, airports were just open fields, and pilots could land in whichever direction they chose. In the late 1930s, because of the increasing weight of aircraft, most airports began to construct concrete runways to replace cinder landing surfaces. These runways were usually about 5,000 feet long and 100 feet wide. In an effort to assist pilots while landing at night or in bad weather, it was eventually determined that runway edge lights should be the standard type of runway lighting.

Runway edge lights are placed on either side of the runway, spaced approximately 200 feet apart, outlining the edges of the runway. These lights are usually placed on short metal poles to elevate them from any obstruction such as long grass or drifting snow. Runway lights are white and are usually covered with a Fresnel lens (see Figure 2–24). Fresnel lenses are designed to focus the emitted light, concentrating it along and slightly above the horizontal plane of the runway's surface.

The lights installed on the last 2,000 feet of runways used for instrument approaches use lenses that are half white and half amber. These lights appear amber to a landing pilot, warning that the far end of the runway is fast approaching. The ends of the runway are clearly designated through the use of runway threshold lights, which are similar to runway lights but use red and green split lenses. As the pilot approaches the runway to land, the threshold lights on the near end of the runway appear green, while those on the far end of the runway appear red.

Figure 2-24. Typical runway light fixture

Runway light systems are normally operated from the control tower and are turned on during nighttime hours and during daylight whenever the visibility is less than 2 miles or at the pilot's request. Whenever the control tower is not in operation, the lights are either left on or are operated using pilot--controlled lighting (PCL) systems.

Runway Light Intensity

Runway light systems are classified according to the brightness they are capable of producing. Low-intensity runway lighting (LIRL) is the least expensive to install and is typically equipped with bulbs that operate at one intensity level. This intensity level is known as step one.

The standard type of lighting for a runway used for instrument approaches is medium-intensity runway lighting (MIRL). Medium-intensity lights are similar in construction to low-intensity lights but are usually equipped with brighter lights. MIRL can be operated on three intensity levels: step one (LIRL equivalent), step two, and step three. Runways that are heavily used during periods of low visibility may be equipped with high-intensity runway lighting (HIRL). High-intensity runway lights operate on five steps. High-intensity lights are operated on step one until the visibility begins to decrease below 5 miles. At that point, higher intensities are used, with step five being reserved for periods when the visibility is less than 1 mile.

Embedded-in-Runway Lighting

Runways that are used extensively during periods of low visibility may be equipped with an assortment of embedded runway lights that provide the pilot additional visual cues when landing. These systems include touchdown zone lighting, runway centerline lighting, and taxiway turnoff lighting.

In conditions of reduced visibility, runway edge lights do not provide sufficient directional information to enable pilots to accurately steer their aircraft along the center of the runway. To assist the pilot, many airports have installed runway centerline lights. Centerline lights are embedded in the runway and are placed along the entire centerline, at 75-foot intervals. Runway centerline lights are bidirectional, so they can be seen by pilots landing from either direction.

For the first part of the runway, the runway is equipped with white centerline lights. The last 1,000 feet of centerline lights are red. Additionally, in the 2,000 feet or runway preceding the red lights, the centerline lights alternate red and white to warn pilots that the runway end is approaching. Runway centerline lights are also varied in intensity in proportion to the setting chosen for the runway edge lights.

Once low visibility landings became commonplace it was apparent that additional visual cues during this critical phase of landing needed to be provided to the pilot. Thus, a new supplemental lighting system was developed, known as touchdown zone lighting. Touchdown zone lights are embedded in the runway on both sides of the centerline and extend from the landing threshold to a point 3,000 feet down the runway. Touchdown zone light intensities are stepped in conjunction with the runway edge lights.

When visibility is reduced, many pilots find it difficult to identify the intersecting taxiways for exiting the runway. Runway utilization rates are reduced as pilots taxi slowly, trying to find the proper turnoff. To reduce this taxi time, some airports have installed taxiway turnoff lights, which are similar to centerline lights but are used to delineate the path that the pilot should use for exiting the runway. Taxiway turnoff lights are inset into the runway's surface and are spaced at 50-foot intervals. These lights are colored green and extend from the runway centerline to the proper intersecting taxiway.

Taxiway Lighting

Large airports may have a myriad of taxiways, runways, and vehicular paths that all look similar to a pilot unfamiliar with the airport. To assist these pilots, taxiway edge lighting systems have been developed. Taxiway edge lights are similar to runway edge lights but operate at reduced wattage and are equipped with blue lenses. Taxiway centerline lights are embedded in the pavement and are green. Taxiway lights may operate at different intensity levels and are usually operated from the control tower.

Approach Lighting Systems

One of the most complex tasks facing pilots occurs near the end of an instrument approach, when they make the transition from instrument to visual flying. During this transition, they must locate the runway and properly maneuver the aircraft for landing within seconds. In conditions of low visibility, a pilot may be able to see only about 2,000 feet ahead of the aircraft. In today's modern jets, this distance can be covered in less than 20 seconds. Within this short time, the pilot must locate the runway, determine the aircraft's position, make any necessary adjustments in speed and flight attitude, and then land the aircraft. Without some form of visual assistance, this task is virtually impossible to perform safely in so short a time.

A number of different types of approach lighting systems have been developed that help the pilot with this critical task. In general, approach lights are placed along the extended centerline of the runway and usually extend from the runway threshold out to a point where the pilot makes the transition from instrument to visual flying. Approach lighting systems are designed to provide the pilot with visual cues that permit accurate aircraft control during the final approach and landing phase of the flight.

The basic configuration of an approach lighting system consists of a series of white lights placed five abreast, extending from the runway threshold out to a distance of about 2,400 to 3,000 feet. These light bars are spaced 100 feet apart. At a point 1,000 feet from the end of the runway, a triple set of light bars is installed to provide the pilot with both roll guidance and a definite, unmistakable distance indication. The threshold of the runway is as delineated with a series of four red light bars and a continuous line of green threshold lights.

To help the pilot pick out the approach light system when transitioning from instrument flying, a high intensity strobe light might be placed on each of the light bars that extended beyond the 1,000-foot mark. These strobe lights flash in sequence, at a rate of two times per second, and appear to the pilot as a moving ball of light leading to the runway. These sequenced flashing lights (SFL) are also referred to by the slang name "the rabbit." This combination approach lighting system became the standard for runways equipped with Category I ILS and is known as approach lighting system type 1 or ALSF-1.

When Category II and III ILSs were being developed, it was realized that an improved approach lighting system was necessary. During Category II approaches, the pilot may be required to transition to visual references during the last 5-10 seconds of flight. Category III approaches permit the pilot even less time to make this transition. In response, the FAA developed an improved approach lighting system known as approach lighting system type 2, or ALSF-2.

ALSF-2 is similar to ALSF-1 but includes additional lighting during the last 1,000 feet nearest to the runway. A supplemental set of white light bars is located 500 feet from the runway threshold to provide the pilot with an additional distance indication. Red light bars are also placed on both sides of the centerline, providing pilots with aircraft roll guidance during the last 1,000 feet.

Both the ALSF-1 and the ALSF-2 systems are expensive to install, operate, and maintain. This expense can be justified only at airports that use this type of equipment routinely. At most airports, a smaller, less expensive system can be used to provide pilots with the same benefits as these larger systems. At these locations, a full-length (3,000-foot) approach lighting system may be unnecessary. For many of these airports, the FAA has chosen to install a version of ALSF-1 that is only 1,200 feet long. This system utilizes the same high-intensity white approach lights as the ALSF-1 system, but they are spaced at 200-foot intervals. This is known as the simplified short approach lighting system (SSALS).

In most of these locations, runway alignment indicator lights (similar to stand alone sequenced flashers) are also installed out to a distance of 2,400 feet. In this configuration, the system is known as the simplified short approach lighting system with RAIL (SSALR). Most ALSF-1 and ALSF-2 systems are wired such that they can operate as SSALR systems during periods when low-visibility approaches are not conducted.

The FAA has begun to place a smaller approach lighting system at airports that do not routinely conduct a large number of low-visibility approaches. This system is designed to include most of the important components available in the ALSF and SSALR systems but reduces the installation, operating, and maintenance expenses even further. This system, known as the medium-intensity approach lighting system with RAIL (MALSR), operates with only medium intensity white lamps. MALSR systems extend 2,400 feet from the runway threshold, with the light bars spaced at 200-foot intervals. MALSR is now the U.S. standard for precision approach lighting. See figure 2-25 for examples of approach light systems.

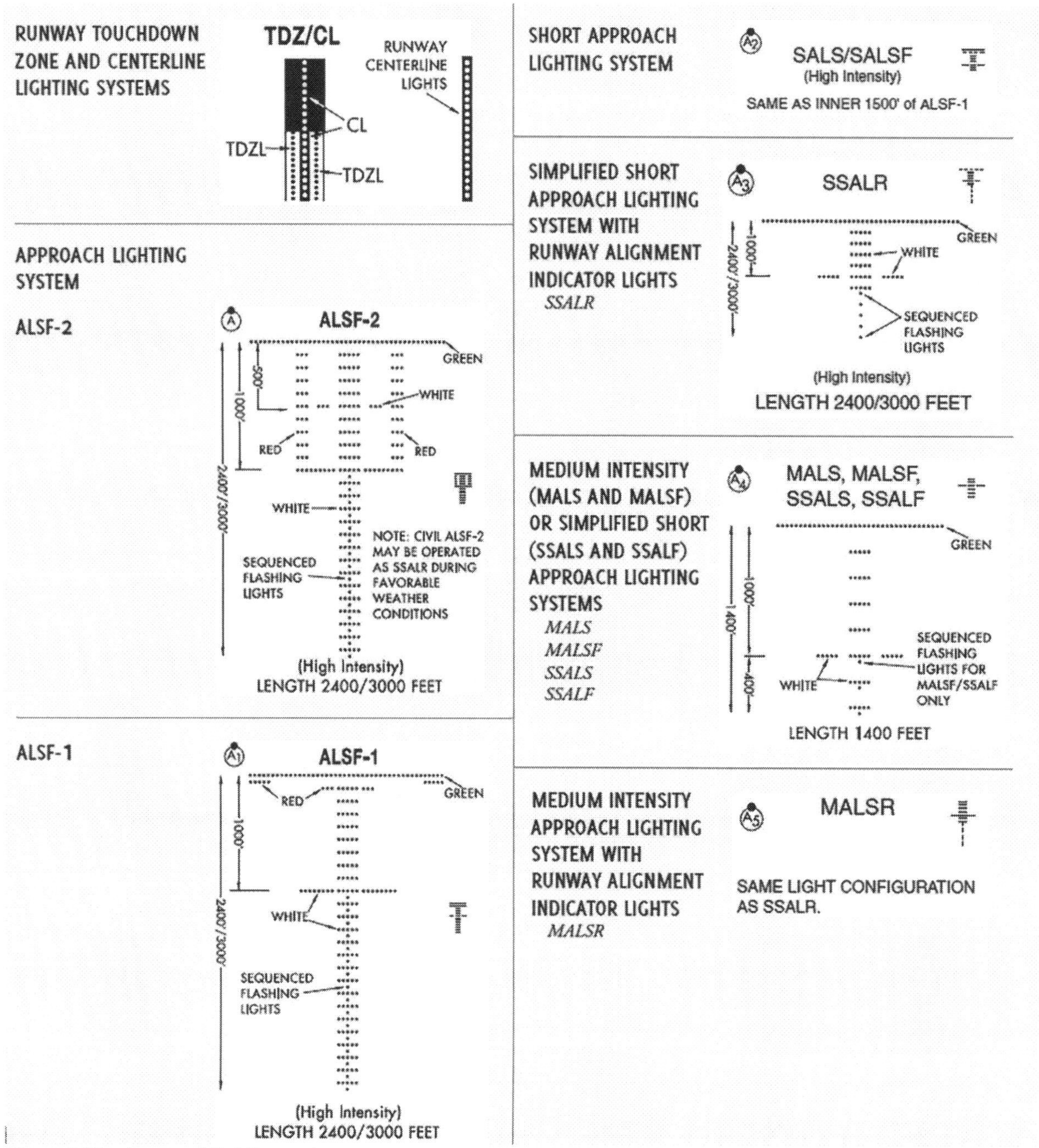

Figure 2-25. Approach lighting systems

Vertical Guidance Systems

Approach lighting systems primarily provide lateral guidance to the pilot, with vertical guidance either being provided by an electronic glide path or through the pilot's judgment. At night, or during periods of reduced visibility, pilots are deprived of many of the visual cues used to determine the proper glide path. Without these cues, pilots may be unable to correctly orient their aircraft during the final approach phase and may misjudge their distance, glide angle, or rate of descent. In 1960, the FAA introduced the visual approach slope indicator (VASI) system designed to provide pilots with accurate glide path information as far as 20 miles from the runway. The VASI system uses multiple light units arranged to provide the pilot with a visual glide path. These light units are next to the runway, with the first located approximately 700 feet and the second approximately 1,200 feet from the approach end.

Each VASI unit provides a narrow beam of light filtered such that the upper portion (above the glide path) of the beam is white and the lower portion (below the glide path) is red. Pilots looking at a VASI light know that the aircraft is too high if they see a white light and too low if they see a red light. VASI units are installed such that a pilot on the desired glide path is above the near VASI (the white beam) but below the far VASI (the red beam). A pilot who is too high will see the white light from both units, whereas the pilot who is too low will see the red beams from both (see Figure 2-26). VASI comes in three different configurations with the only difference being the number of light units installed and their placement next to the runway.

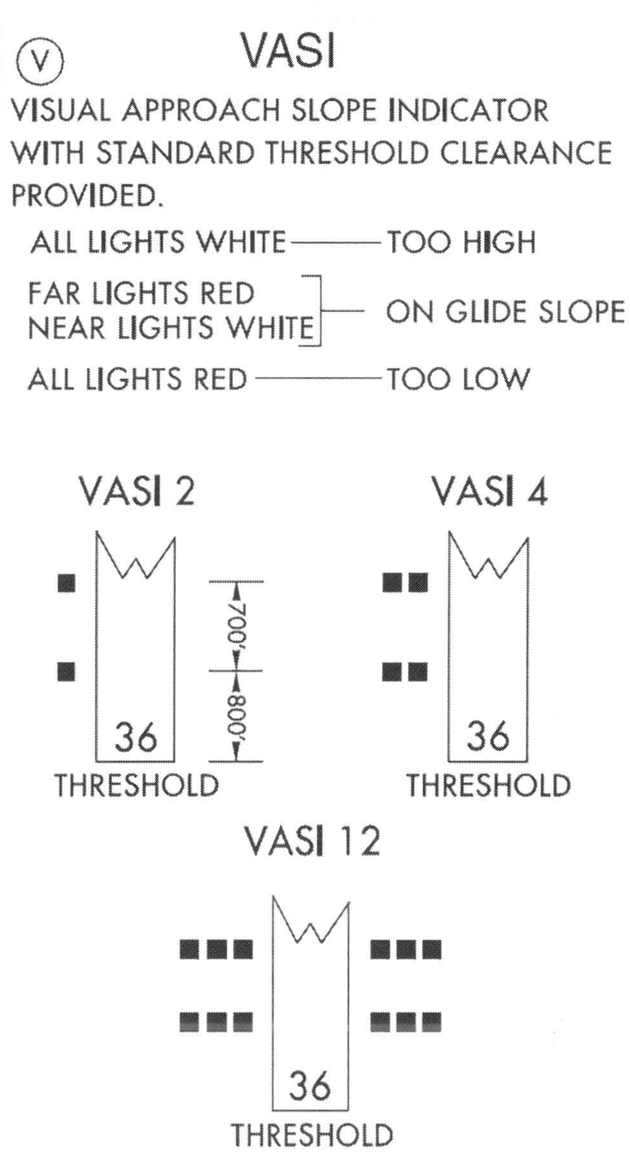

Figure 2-26. VASI installation

Precision Approach Path Indicator

VASI is highly effective but can be difficult to use since the pilot must constantly observe light units that are separated by up to 1,000 feet. A similar system, the precision approach path indicator (PAPI), has been developed that remedies this situation. PAPI is similar to VASI but is installed in a single row. Each light unit emits a white and a red beam but at progressively higher angles. If the pilot is more than half a degree above the desired flight path, all the light units will appear to emit white light. But as the pilot descends to a lower angle, the system is designed so that the pilot will begin to see red light emitted from the unit nearest to the runway. (See figure 2-27).

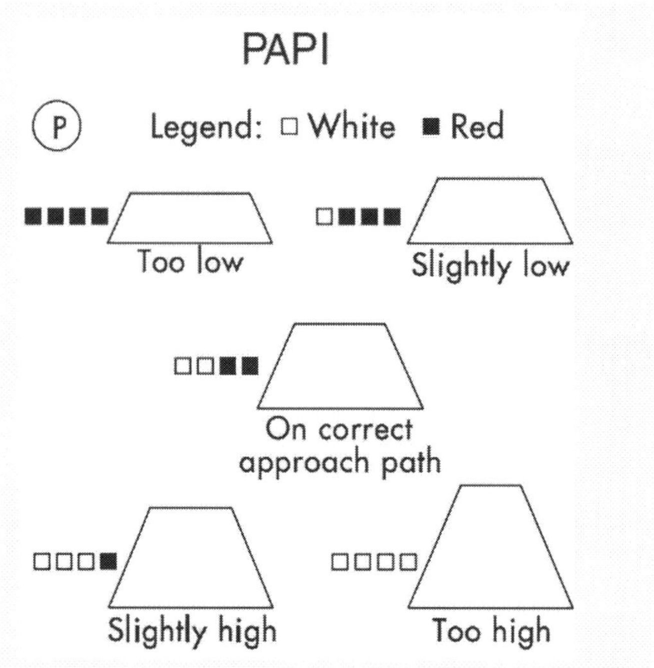

Figure 2-27. PAPI system

When half the lights are red and the other half are white, the pilot is on the desired glide path, which is usually 3°. If the pilot descends below this glide path angle, additional light units will be observed as red. If all the light units appear red, the pilot is in excess of half a degree below the desired glide path and should begin to climb immediately.

Chapter 3

Air Traffic Control System

Airspace

To provide flexibility while still insuring appropriate levels of safety, United States airspace has been categorized into different classes, with specific requirements and differing rules for operating in each class. Different airspace classifications and rules permit the FAA and other national agencies to provide varying levels of security and control as needed. In general, airspace classification is designed to provide maximum separation and active control in areas of dense or high-speed flight operations while affording as much flexibility as possible.

The requirements for operating in each airspace area as well as the ATC services offered differ based on whether the aircraft is operating under visual flight rules (VFR) or instrument flight rules (IFR). Under IFR, the ATC system separate pilots from other participating aircraft. IFR aircraft are required to file flight plans, receive operating clearances, maintain contact with ATC, and adhere to ATC instructions. Airline, military, most corporate and many general aviation flights are conducted under IFR. IFR flights occur in both IFR and VFR weather conditions; in almost every case, IFR aircraft are separated from other IFR aircraft by ATC. However, when they are below 18,000 feet, IFR pilots normally still need to keep a look out for VFR aircraft operating nearby. VFR weather specifications determine the rules by which aircraft can operate without using ATC to provide aircraft separation. Pilots operating VFR provide their own separation using the principle of "see and avoid". VFR is primarily used by smaller and slower aircraft and is limited to areas without a lot of air traffic. There are specific weather conditions that must exist to fly VFR (known as VFR weather minima) and pilots can only fly VFR when those weather conditions exist.

Airspace Classifications

To more easily define the air traffic control services that can be offered as well as the weather minima required for VFR and IFR options, the airspace above the U.S. has been classified into one of three general categories.

- **Controlled airspace** - ATC separates IFR aircraft but VFR pilots provide their own separation (in some airspace classes, ATC separates every type of aircraft. This is known as "positive control" and is conducted in positive controlled airspace).
- **Uncontrolled airspace** - pilots provide their own separation, regardless of whether they are IFR or VFR.
- **Special use airspace (SUAS)** - airspace within which there are special operating restrictions and rules.

Airspace Classes

To help differentiate the services offered in controlled and uncontrolled airspace, the FAA, as well as the International Civil Aviation Organization (ICAO), have developed seven standardized airspace categories, defined as Class A, B, C, D, E, F and G. (See figure 3-1). Classes A through E are designated as controlled airspace, Class F is not used in the U.S, and Class G airspace is uncontrolled. In general, Class A airspace is the most restrictive, where all aircraft are required to operate IFR (positive controlled). Class G airspace is the least restrictive with few or no ATC services provided. Class B, C, D, and E airspace span the range of services, with a mixture of IFR and VFR aircraft. Special use airspace is specific to the U.S. and does not follow specific ICAO classifications; it is used primarily by military aircraft or in protection of classified or security areas, or for routine training.

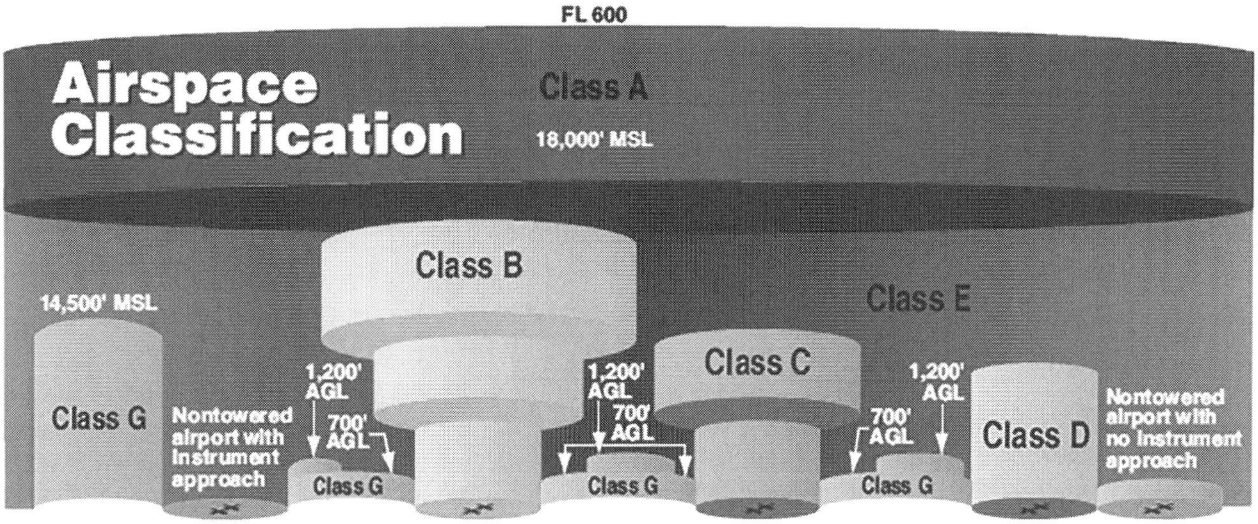

Figure 3-1. Airspace classes in the U.S.

IFR Flight in Controlled Airspace (Class A, B, C, D, and E)

Within controlled airspace, air traffic controllers are always required to separate IFR aircraft and are sometimes required to separate VFR aircraft using the procedures specified in the Air Traffic Control Handbook Order 7110.65. VFR aircraft can operate in some areas of controlled airspace; in class E they can do so without even contacting ATC, therefore it remains the responsibility of IFR pilots to see and avoid these aircraft, regardless of the services being provided by the air traffic controller.

Aircraft Separation in Controlled Airspace

The FAA has divided the nation's airspace into twenty-two areas and assigned aircraft separation responsibility within these areas to the air route traffic control centers mentioned in chapter one. The basic function of an ARTCC is to separate IFR aircraft traveling between airports. If, after performing a study of the local airspace and traffic structure, the FAA determines that both safety and efficiency would be increased if a local facility (control tower or TRACON), were to take responsibility for some of their airspace, the ARTCC delegates aircraft separation responsibility to that facility through a letter of agreement (LOA). Such a letter between air traffic control facilities specifically clarifies the following:

- The physical dimensions of the airspace involved.
- The approved altitudes and airways used by aircraft that cross the boundary between the two facilities.
- The procedures used by air traffic controllers when an aircraft progresses from one facility's area of responsibility into the next.
- The separation responsibilities of each facility.
- The communication procedures to be utilized to coordinate information.

Air Traffic Control Procedures

Controllers must use the procedures found in the Air Traffic Control Handbook (JO 7110.65) when separating aircraft. In many ways, this handbook is similar to separation and procedural guidelines published by ICAO; it does, however, differ in some areas. FAA-certified air traffic controllers, whether working for the FAA or for another employer, are obligated to use the handbook procedures whenever they are performing air traffic control duties. If the 7110.65 differs from ICAO rules, the rules in the 7110.65 have precedence for operations within the United States.

The FAA has the responsibility of separating every aircraft participating in the nation's air traffic control system. A participating aircraft is defined as the following:

- Any aircraft operating under an FAA clearance in controlled airspace (classes A, B, C, D or E) and using IFR flight rules.
- VFR aircraft operating within areas of designated airspace where air traffic control participation is mandatory (certain operations in Class B, C, or D airspace).

Sectors

As a single controller cannot possibly separate every aircraft within a particular facility's boundaries, every ARTCC and/or TRACON is divided into numerous smaller areas called sectors. Each sector may have a single controller or a team of controllers assigned to it at any given time. Each sector is fashioned in a logical manner, taking into consideration the airway structure and traffic flows. Every ARTCC's airspace is partitioned both vertically and horizontally into between twenty and eighty sectors. (See figure 3-2). The airspace at most centers is usually stratified into at least two levels: a low-altitude group of sectors extending from the earth's surface up to 24,000 feet above mean sea level (MSL) and a high-altitude group of sectors extending from 24,000 feet to 60,000 MSL; some centers further stratify the upper airspace. In general, aircraft operating on low-altitude airways (victor airways) are always separated by low-altitude controllers, whereas aircraft operating on high-altitude jet routes are separated by high-altitude controllers. All controllers apply the 7110.65 separation rules to all aircraft, so aircraft should not notice any difference in procedures between different sectors.

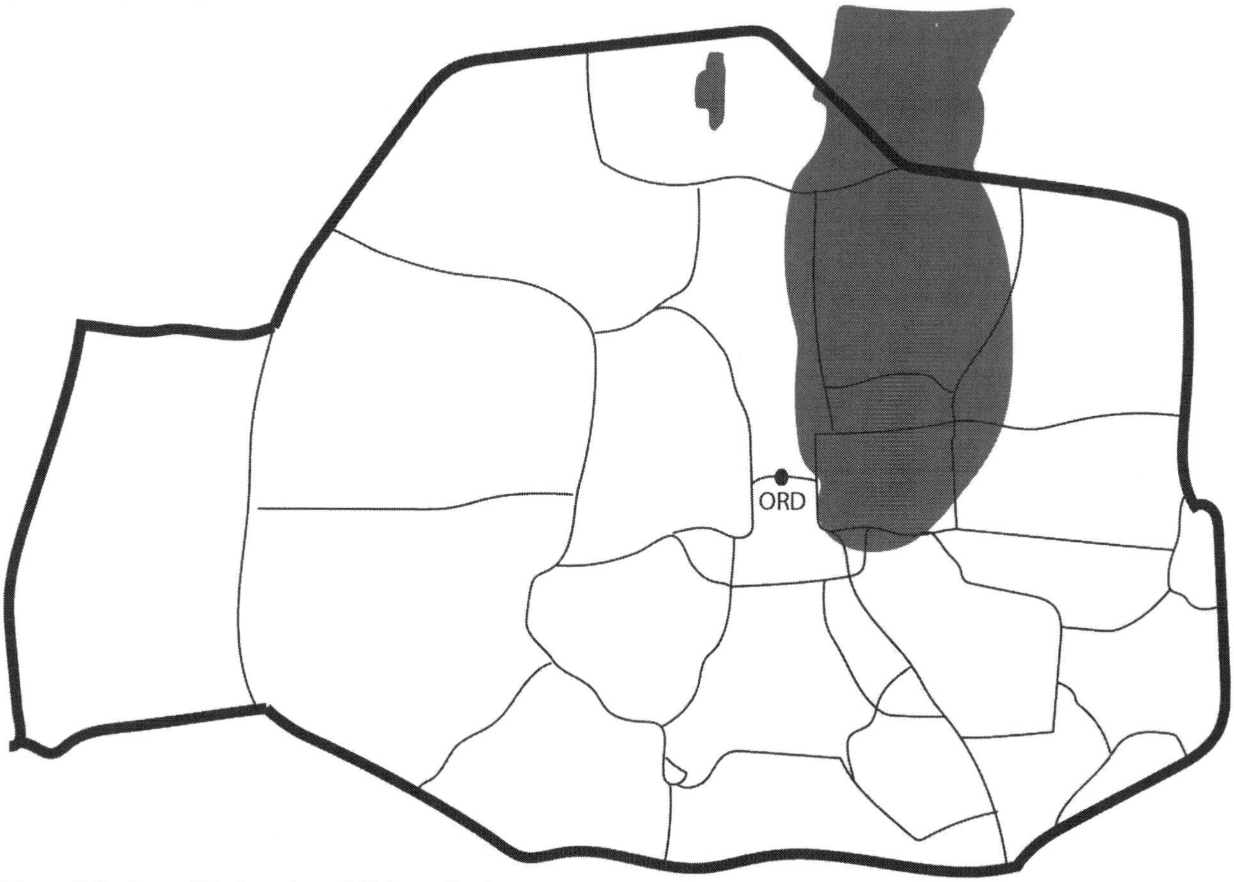

Figure 3-2. Low altitude sectors at Chicago Center

Hand Off Procedures

When an aircraft crosses a sector boundary, the responsibility for separating that aircraft passes on to the controller in the new sector. The original controller is known as the transferring controller, whereas the next controller is called the receiving controller. This transfer of separation responsibility is known as the transfer of control, or more commonly as a handoff. Typically, the pilot is directed to contact the receiving controller on a different radio frequency prior to crossing the sector boundary. This is known as the transfer of communication, or more commonly as a frequency change. This should not take place until the transfer of control has been accomplished. (See figure 3-3)

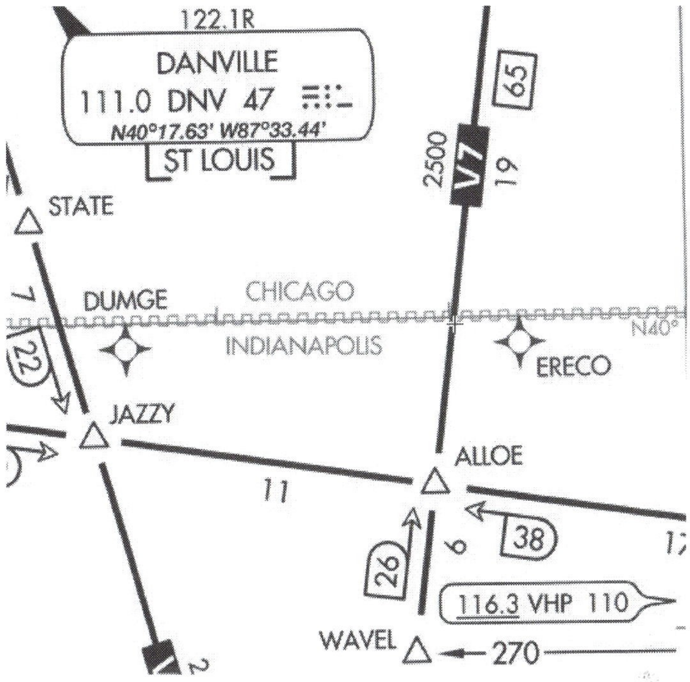

Figure 3-3. Facility boundaries

Handoffs are necessary when aircraft cross sector boundaries and when an aircraft crosses the boundary between two separate ATC facilities, such as between two centers or between a tower and a center. The FAA handbook specifies that both the transfer of control and communication must occur before the aircraft crosses the boundary. This ensures that the receiving controller will be able to evaluate how the aircraft integrates with their existing traffic, as well as to be in radio contact with the pilot before the aircraft enters his or her sector. This does not permit the receiving controller to issue any new control instructions to the pilot before the aircraft crosses the sector boundary unless they have coordinated with the previous controller. Doing so without prior coordination is considered an operational deviation. Likewise, once the transferring controller has completed a handoff, they are not allowed to make any changes to route, altitude, or anything else that would affect an aircraft's progress; doing so would also be considered a deviation.

Controller Duties in an Air Route Traffic Control Center

Flight Data Controllers

Every sector within an ARTCC usually has one to three controllers assigned to separate the aircraft within that sector. The first position that most controllers in an ARTCC are assigned is that of flight data controller. The flight data controller is responsible for assisting other controllers who actually separate the aircraft. The flight data controller effects coordination with other controllers and passes along pertinent flight information to controllers working in other sectors. This position is frequently called the A-side. One of the primary functions of this position previously was to deliver flight progress strips to various sectors. As most centers have moved to electronic strips, this position is no longer commonly staffed.

Radar Controllers

Every ARTCC sector equipped with radar is staffed by a controller whose responsibility is to separate aircraft using radar procedures. Radar controllers issue altitude, heading, or airspeed changes to keep aircraft separated and in compliance with the various letters of agreement and facility directives that may apply to that sector (See figure 3-3). They are also in charge of initiating and receiving handoffs. Additionally, they are the controller who primarily speaks with the sector's aircraft; as such, this position is always staffed for each sector. This position is commonly known as the R-side. If sectors are not busy, more than one sector can be combined to this position, so that one controller can serve this function for two or more sectors in times of low traffic.

Radar Associate/Nonradar Controller

Depending on how busy a sector is, each sector may also be staffed by a radar associate/nonradar controller whose duties are to assist the radar controller when separating aircraft that do not appear on the radar display. The nonradar controller's duties include ensuring flight progress strips accurately reflect every aircraft's position, altitude, and route of flight. The nonradar controller uses this information to separate aircraft that are either too low or too far away to be displayed on the radar. The nonradar controller must be prepared to assume aircraft separation responsibility if the radar display should malfunction. Additionally, the radar associate is responsible for ensuring that all coordination is completed with any other sectors or facilities, as well as issuing clearances to aircraft departing airports not served by a control tower. This position is also known as the D-side.

Figure 3-4. Controllers working at an ARTCC

Air Traffic Control Tower Responsibilities

Flight Data

The first position that most new controllers in a tower will be assigned to is that of the flight data controller. Just as in a center, the flight data controller is responsible for assisting other controllers who actually separate aircraft. The flight data controller might also have the responsibility for recording the automatic terminal information service (ATIS) that provides pilots with airport weather conditions, runways in use and other important airport information. This position is frequently combined with the clearance delivery position.

Clearance Delivery

The clearance delivery controller issues IFR clearances to aircraft. If the aircraft is headed to a busy or weather-constrained airport, clearance delivery coordinates with either the ARTCC or system command center to obtain releases for each aircraft. Additionally, all IFR aircraft departing VFR towers need an IFR clearance from the overlying radar facility, which the clearance delivery position will receive and relay. The clearance issued does not allow the aircraft to depart or move around the airport; it only authorizes aircraft to operate in the national airspace system under certain conditions after departure. The ground and local controllers are still responsible for ensuring separation until the aircraft departs.

Ground Control

Ground Control is responsible for the separation of aircraft and vehicles operating on the ramp, taxiways, and inactive runways. This responsibility includes aircraft taxiing for takeoff or to parking after landing, as well as any ground vehicles operating on airport movement areas. Airport movement areas do not include those areas solely reserved for vehicular traffic such as service roads or boarding areas. The ground controller is assigned a unique radio frequency to communicate with pilots and vehicle operators. The most common ground control frequency is 121.90 MHz. (See figure 3-4).

Local Control

The local controller is primarily responsible for the separation of aircraft operating within the airport traffic area and those landing and departing on any of the active runways. The local controller is assigned a unique radio frequency for communicating with these aircraft. The primary responsibility of the local controller is arranging inbound aircraft into a smooth and orderly flow of traffic and sequencing departing aircraft into this flow. The local controller's responsibilities are complicated by the fact that most airports do not have sufficient nonintersecting runways to handle the number of aircraft that want to land or take off. Thus, the local controller may be forced to use two or three runways that intersect each other. At very busy facilities, the local controller's workload may be too much for one person to handle. In that case, the local control position is split into two, with each controller responsible for different runways and assigned separate radio frequencies.

Figure 3-5. Controllers working in a control tower

Approach and Departure Control

At busy facilities that have been delegated a large amount of airspace, an approach and departure control position may be designated. This position (or positions) are commonly referred to simply as approach control. At smaller, less busy towers, approach control may be the responsibility of one controller stationed in the tower cab itself, but at larger and busier airports, the approach control may be housed in a separate building located near the tower. This facility is known as a terminal radar approach control, or TRACON.

A TRACON may be equipped with up to twenty radar displays and may be staffed by up to forty controllers at a time. (See figure 3-6). At some facilities, TRACON controllers occasionally work in the tower cab, but at others they are assigned strictly to the approach control facility. The airspace controlled by a TRACON is usually too large to be administered by one controller and, like the center, is divided into smaller, more manageable sectors. In recent years, a trend toward larger, consolidated TRACON facilities that house the approach control facilities for many airports has emerged. These "super" TRACONS cover large geographical areas and can be staffed by several hundred controllers, all of whom will be divided into their airport specialties.

Figure 3-6. TRACON controllers

Flight Operations

Aircraft Flight Plan

Before beginning IFR flight in controlled airspace, the pilot is required to file a flight plan with the FAA and receive a clearance from an ATC facility. A general aviation or corporate pilot usually files an IFR flight plan with a flight service station, using the internet or telephone, and the information is then forwarded to the ARTCC with jurisdiction over the departure airport. Airline flight plans are typically filed directly with the FAA using stored flight plan information or through flight dispatch. If a pilot needs to file a flight plan while airborne, the ATC facility in contact with the pilot transmits the flight plan information to the proper ATC facility.

The information required on a flight plan includes:

- Type of flight plan, VFR, IFR, or DVFR (which is a special type of VFR flight plan used if an aircraft is entering, leaving, or transiting U.S. airspace).
- Aircraft identification number. This is either the aircraft's assigned serial number, if it is a general aviation, military or corporate flight, or the airline call sign and flight number.
- Aircraft type and navigation equipment installed on the aircraft. The aircraft type is abbreviated, using codes found in the FAA Contractions publication (JO 7340.2). Examples of aircraft type identifiers include the following:

TABLE 3-1 EXAMPLES OF AIRCRAFT TYPE IDENTIFIERS

Aircraft	FAA Identifier
Airbus A-320	A320
Boeing 737-800	B738
Cessna Citation 650	C650
Cessna 172 Skyhawk	C172
Cirrus SR-20	SR20
Diamond DA-42	DA42
Embraer EMB-190	E190
Gates Learjet 55	LJ55
General Dynamics F16 Falcon	F16
Hawker Beech 200 King Air	BE20
Piper PA-28 Warrior	P28A

  - The pilot must also identify the navigational capabilities of the aircraft by appending a unique suffix code to the aircraft type. The aircraft type is separated from the equipment code with a slash. Some of the more common equipment codes include:

TABLE 3-2 COMMON EQUIPMENT CODES

Suffix	Equipment Capability
/A	VOR and DME equipped
/L	GNSS equipped

- The aircraft's cruising true airspeed in knots.
- The abbreviation for the departure point. This is normally the departure airport but can be an enroute fix.
- The proposed time of departure.
- The pilot's requested cruising altitude.
- The requested route of flight. This must include the airway and navigation aid identifiers. The entire route of flight must be specified. When changing from one airway to another, the intersection fix is always specified. If no airway is being used, only the navaids used need to be specified. For example, the route ALB J37 BUMPY J14 BHM would be interpreted as departing Albany, New York, flying via Jet Route 37 until reaching the BUMPY intersection, transitioning to Jet Route 14 at BUMPY intersection, then flying via Jet Route 14 to the destination which is Birmingham, Alabama. The route ALB BHM would be interpreted as departing Albany and flying in a straight line (direct) to the destination airport at Birmingham.
- The destination airport.
- The estimated time enroute in hours and minutes.
- Any pertinent remarks.
- Fuel on board the aircraft expressed in hours and minutes.
- The pilots selected alternate airport.
- Name and address of the pilot in command.
- Number of people on board.
- Color of the aircraft.
- Contact information at destination airport.

ATC Clearance

After the flight plan is filed and processed, an air traffic clearance is generated and transmitted to the pilot at the time they request the clearance. Clearances are never given early; they are based upon existing traffic conditions at the time of request. Filing the flight plan provides the ATC facilities with valuable time-saving information; it is not considered an official request by the pilot until verbally confirmed. The clearance includes the destination, departure procedures, route of flight, altitude and any other related information. The aircraft is now authorized to operate in the national airspace system (using IFR flight rules) but are subject to control instructions issued by various air traffic controllers. The specific ATC services provided may differ slightly depending on the airspace class being operated in, but in general ATC will now separate IFR aircraft from all other aircraft operating IFR.

ATC Services in Different Airspace Classes

Class A

Class A airspace general exists from 18,000 feet to 60,000 feet; most commercial traffic travels in this altitude stratum. The federal aviation regulations require that every aircraft operating within Class A airspace operate under instrument flight rules and receive a clearance from ATC. VFR flight is not permitted in Class A airspace. Class A airspace is not designated on navigational charts.

Class B

Class B airspace exists around the nation's busiest airports up to an altitude of about 10,000'. (See figure 3-7). The separation procedures applied to aircraft operating within Class B airspace are similar to those applied to aircraft operating in Class A airspace. VFR pilots can enter Class B airspace on one of several routes, known as transition areas, flyways, or byways. Most of these require a clearance from ATC before doing so, and some separation will be provided to the pilots on these clearances. One particular feature of Class B airspace is the presence of a Mode C veil, which requires VFR aircraft to have a Mode C transponder (capable of submitting altitude information to ACT facilities) in order to ensure that VFR aircraft are operating below or above the Class B airspace.

The following terminal areas around the country are currently designated as Class B airspace:

TABLE 3-3 CLASS B AIRSPACE

Location
Atlanta, GA
Baltimore, MD-Washington, D.C. area including:
Washington Dulles International Airport
Washington National Ronald Reagan Airport
Baltimore/Washington International Airport
Andrews Air Force Base
Boston, MA
Charlotte, NC
Chicago O'Hare, IL
Cincinnati, OH-(Covington, KY)
Cleveland, OH
Dallas, TX area including:
Dallas/Fort Worth International Airport
Dallas Love Field Airport
Denver, CO
Detroit, MI
Honolulu, HI
Houston, TX area including:
George Bush Intercontinental
William P. Hobby Airport
Kansas City, MO
Las Vegas, NV
Los Angeles, CA
Memphis, TN
Miami, FL
Minneapolis, MN
New Orleans, LA
New York, NY-Newark, NJ area including:
LaGuardia Airport
John F Kennedy International Airport
Newark Liberty International Airport
Orlando, FL
Philadelphia, PA
Phoenix, AZ
Pittsburgh, PA
Saint Louis, MO
Salt Lake City, UT
San Diego, CA
San Francisco, CA
Seattle, WA
Tampa, FL

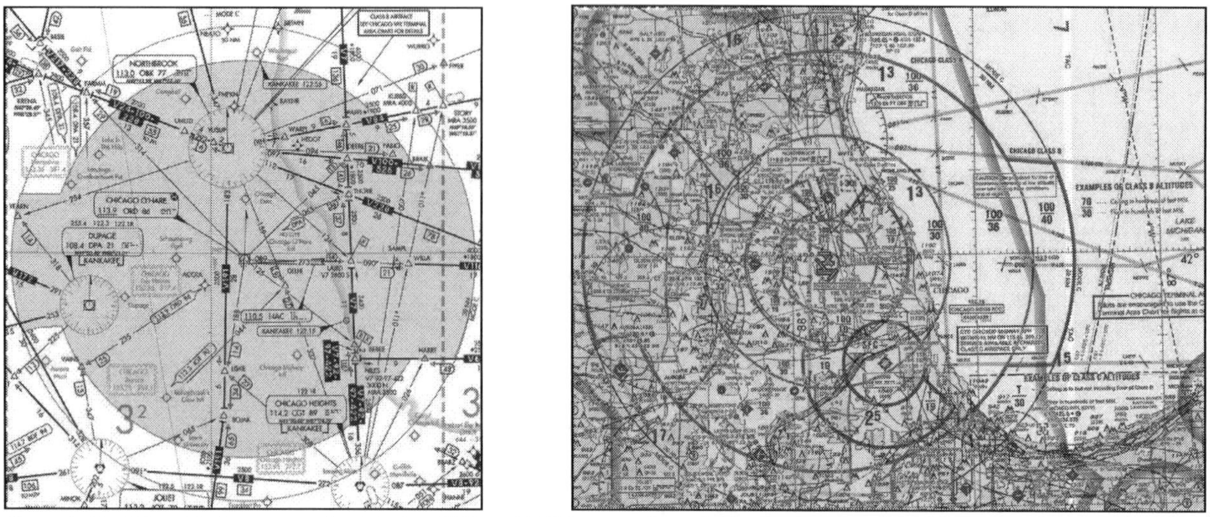

Figure 3-7. Class B airspace as depicted on IFR and VFR navigation charts.

Class C

Class C airspace surrounds medium-activity airports. (See figure 3-8). Class C airspace extends from the Earth's surface, up to about 4,000 feet above ground level. Within Class C airspace, the controller separates all IFR aircraft from other aircraft (either IFR or VFR) and issues safety alerts if two VFR aircraft appear to be too close to one another. VFR pilots do not need a clearance to enter class C airspace; they simply need to make radio contact with ATC. However, if they are departing the primary airport, they do need to be in contact with ATC to receive clearance to taxi and takeoff. VFR aircraft entering the Class C airspace are required to establish communication with ATC prior to entry. Aircraft operating in Class C airspace are restricted to a speed of 200 knots within four miles of the primary airport, up to 4,000 feet. This is to accommodate the numerous, slow-moving VFR aircraft that frequently operate in and near Class C airspace.

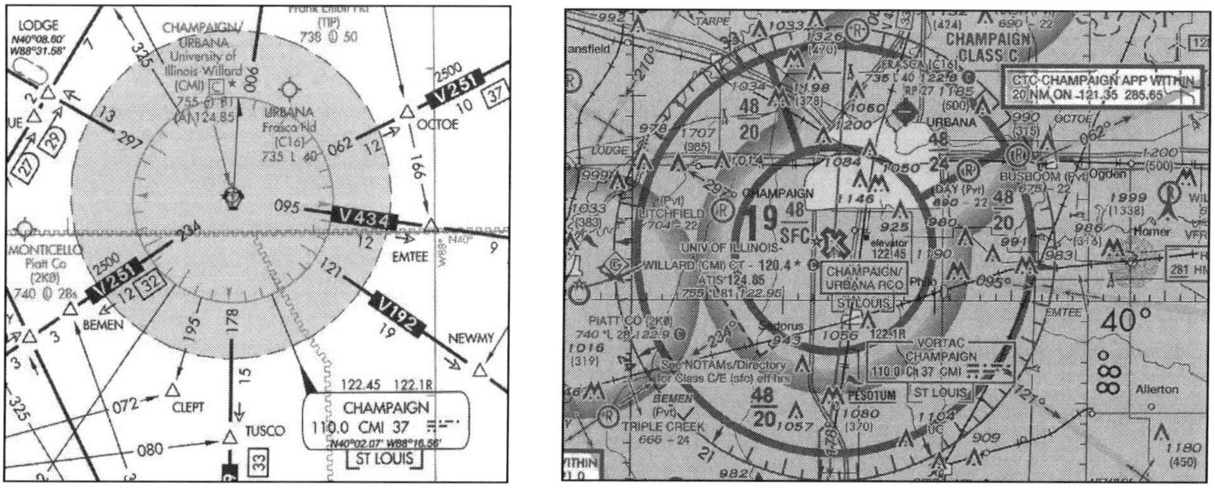

Figure 3-8. Class C airspace as depicted on IFR and VFR navigation charts.

The airports listed on Table 3-4 have been established as Class C airspace airports.

TABLE 3-4 CLASS C AIRSPACE

Location
Alabama – Birmingham, Huntsville, Mobile
Alaska – Anchorage
Arizona – Tucson, Davis Monthan AFB
Arkansas – Little Rock, Fayetteville
California – Beale AFB, Burbank, Fresno, Monterey, Oakland International, Ontario, March Air Reserve Base, Sacramento, Santa Barbara, John Wayne Orange County, San José
Colorado – Colorado Springs
Connecticut – Hartford-Bradley International
Florida – Daytona Beach, Fort Lauderdale-Hollywood, Jacksonville, NAS Whiting Field (North and South), NAS Pensacola, Palm Beach, Pensacola Regional, Southwest Florida-Fort Myers, Orlando-Sanford, Sarasota-Bradenton, Tallahassee
Georgia –Savannah
Hawaii – Kahului-Maui
Idaho – Boise
Illinois –Champaign-Urbana, Chicago Midway, Quad City-Moline, Greater Peoria, Capital-Springfield
Indiana – Evansville, Fort Wayne, Indianapolis, South Bend
Iowa – Cedar Rapids, Des Moines
Kansas – Wichita
Kentucky – Lexington, Louisville-Standiford
Louisiana – Barksdale AFB, Baton Rouge, Lafayette, Shreveport
Maine – Bangor, Portland
Michigan – Flint, Grand Rapids, Lansing
Mississippi – Columbus AFB, Jackson
Missouri – Springfield
Montana – Billings
Nebraska – Lincoln, Offutt AFB, Omaha
Nevada – Reno
New Hampshire – Manchester
New Jersey – Atlantic City
New Mexico – Albuquerque
New York – Albany, Buffalo, Long Island MacArthur, Rochester, Syracuse
North Carolina – Asheville, Fayetteville, Greensboro-Piedmont Triad, Pope AFB, Raleigh-Durham
Ohio – Akron-Canton, Columbus, Dayton, Toledo
Oklahoma – Oklahoma City, Tinker AFB, Tulsa
Oregon – Portland
Pennsylvania – Allentown-Bethlehem-Easton
Rhode Island – Providence
South Carolina – Columbia, Charleston, Greenville-Spartanburg, Myrtle Beach, Shaw AFB
Tennessee – Nashville, Chattanooga, Knoxville
Texas – Abilene, Amarillo, Austin, Corpus Christi, Laughlin AFB, Dyess AFB, El Paso, Harlingen, Lubbock, Midland, San Antonio
Vermont – Burlington
Virginia – Norfolk, Roanoke, Richmond
Washington – Spokane, NAS Whidbey Island, Fairchild AFB, Vancouver
West Virginia – Charleston
Wisconsin – Green Bay, Milwaukee, Madison
Puerto Rico – San Juan

Location
Virgin Islands – St. Thomas

Class D

Class D airspace exists from the earth's surface up to 2,500 feet above the airport elevation. (See figure 3-9). Class D airspace surrounds those airports that have an operating control tower. Similar to Class C airspace, VFR pilots are required to establish two-way radio communication with the air traffic control tower prior to entering the Class D airspace. Even though a control tower within class D airspace might not have radar, they can still separate IFR aircraft from each other; however, they are only able to advise VFR aircraft of possible traffic. Controllers in Class D airspace are responsible for separation on the landing and departing runways, and on the controlled airport surface areas. Also, similar to Class C airspace, pilots operating in Class D airspace are speed restricted to 200 knots below 4,000 feet within 4 miles of the primary airport.

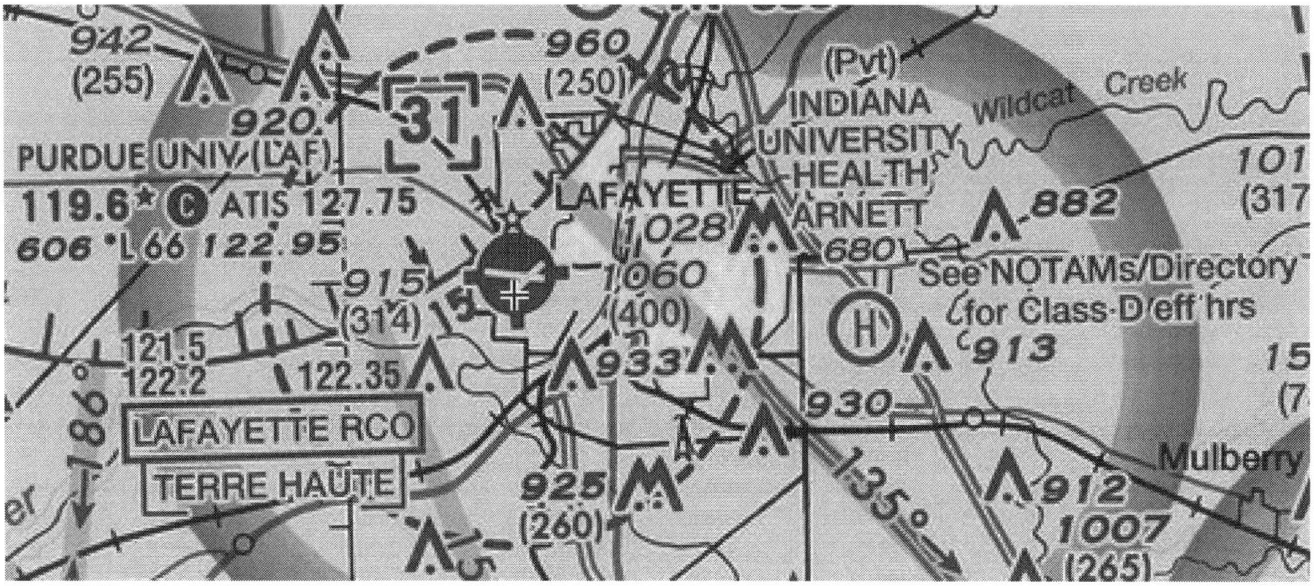

Figure 3-9. Class D airspace as depicted on a VFR navigation chart

Class E

Generally, if the airspace is not Class A, B, C, or D, and not very close to the ground (700' or 1,200' AGL), it will be designated as Class E airspace. In most of the continental U.S., Class E airspace extends down from 18,000' MSL to the top of other designated airspace. All low-altitude federal, or victor airways are Class E airspace. IFR aircraft within class E airspace are separated from other IFR aircraft by ATC. VFR aircraft must "see and avoid" other aircraft (both VFR and IFR). VFR aircraft must operate clear of clouds and with enough visibility to properly see and avoid these other aircraft. ATC can assist IFR aircraft by advising them of known VFR traffic, but since VFR aircraft may not be in contact with ATC (they aren't required to be), ATC may be unaware of their existence.

Class F and G Airspace

Class F airspace is not used in the United States. It is used, however, internationally in areas of limited ATC capability. Class G airspace is used in the U.S. and is uncontrolled airspace within which ATC services are not provided to any aircraft. The regulations for flight in uncontrolled airspace are quite specific and place the burden of separation, whether operating IFR or VFR), on the pilot. Most of the uncontrolled airspace in the US is located away from major airports and below 1,200 feet AGL.

Airways

As mentioned earlier, most of the airways across the country fall within class E airspace, (or A, B, C and D). In the past, there were only VOR-based airways and the route identifiers were based on the altitude of the aircraft. VOR airway numbers are preceded by the letter V (pronounced Victor) if they are low-altitude airways, or the letter J (pronounced Jay) if they are high-altitude airways. In 2004, the FAA began to establish RNAV low altitude routes around or through busy terminal areas. The FAA used the flexibility provided by RNAV to develop a point-to-point route capability for busy terminal and other restrictive areas. Those routes are called Tango routes. This program was later expanded to include high altitude GPS-based routes to and from airports. These new airways are prefixed by the letter "Q" and are referred to as "Q routes". (See figures 3-10 and 3-11).

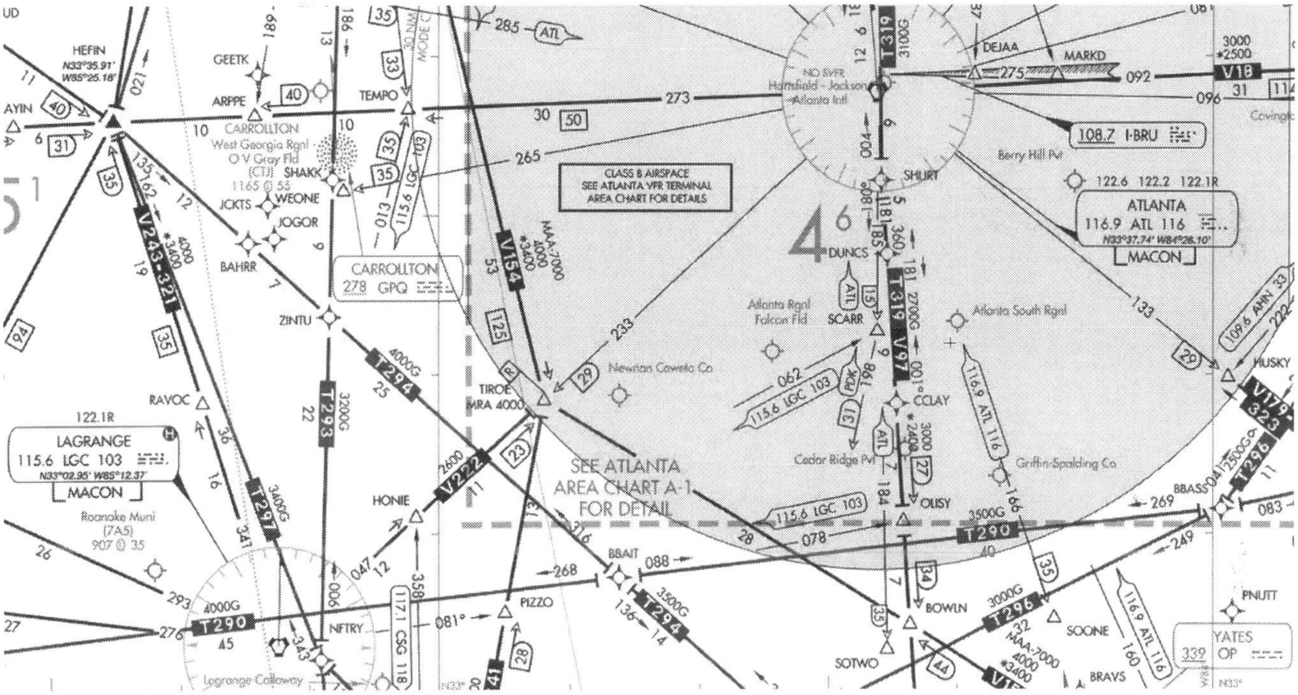

Figure 3-10. V and T routes in the vicinity of Atlanta, Georgia

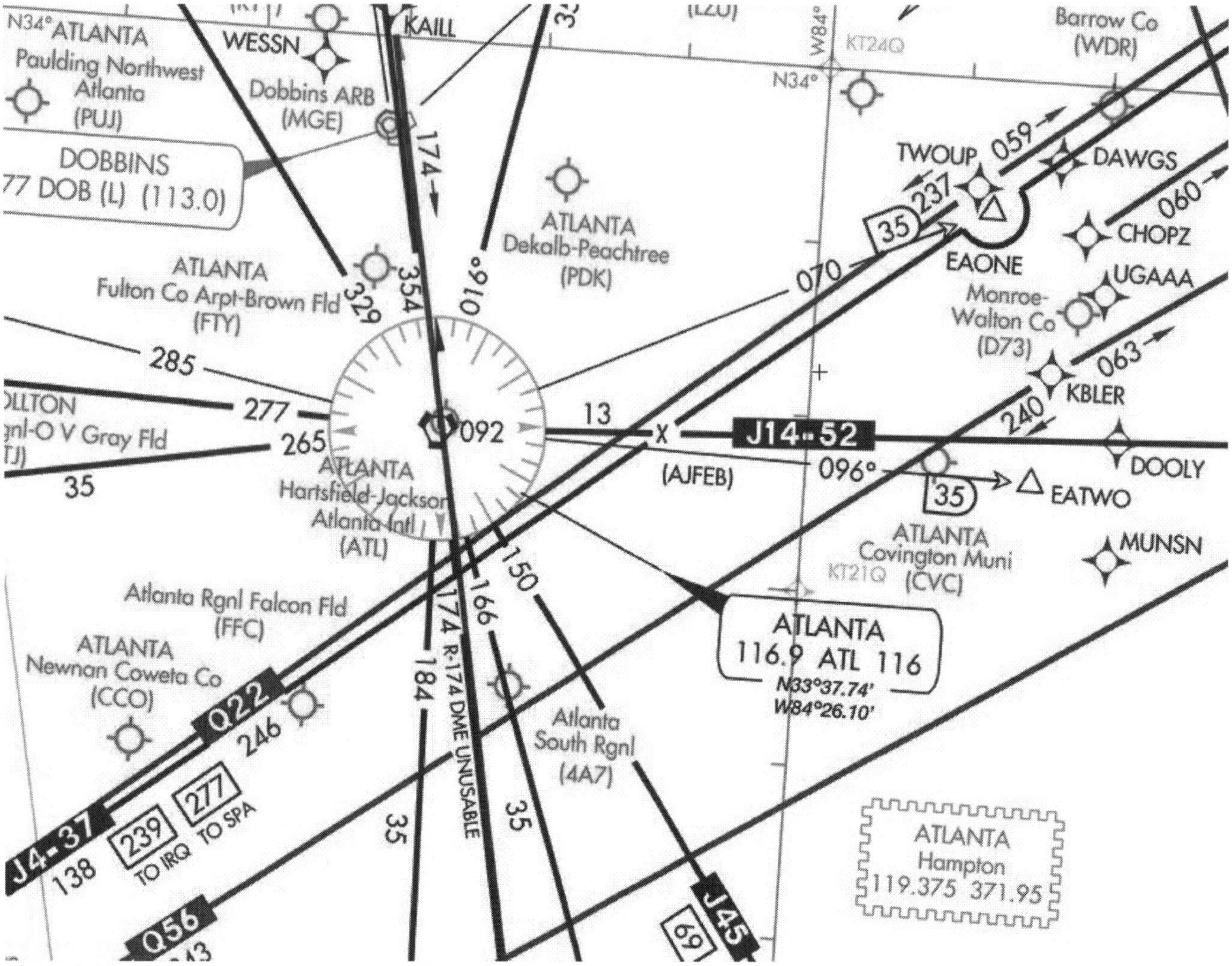

Figure 3-11. J and Q routes in the vicinity of Atlanta, Georgia

Aircraft Altitude and Flight Levels

If an aircraft altimeter is to operate correctly, the local barometric pressure must be manually set by the pilot. Since aircraft using high-altitude airways usually travel at high airspeeds, it is difficult to ensure that every aircraft operating within a given area is using the same altimeter setting. It is important that every aircraft altimeter measure altitude above the same reference plane (MSL). If two aircraft using different altimeter settings were flying in close proximity, they could conceivably be at the same altitude even though their altimeters would indicate different altitudes.

Low-altitude aircraft (below 18,000 MSL) set their altimeter to the current station pressure at the nearest ATC facility; this ensures that every aircraft operating within the same area is using the same altimeter setting.

At high altitudes (18,000' and above) pilots to set their altimeters to 29.92 inches of mercury which is known as standard pressure. To reduce the possibility of a pilot mistakenly using the local altimeter setting when flying on a high-altitude route, any cruising altitude at or above 18,000 feet MSL is referred to as a flight level (FL). Each flight level is stated using three digits that represent hundreds of feet. For example, FL250 represents a barometric altimeter indication of 25,000 feet.

Setting the altimeter to standard pressure when operating at or above this transition level ensures that every aircraft is using the same altimeter setting and measuring altitude from a common datum. The only problem with this procedure is that the altimeter is no longer indicating true altitude, which makes it difficult to determine the aircraft's clearance above a nearby obstruction or another aircraft operating in the vicinity. Although few obstructions occur at these altitudes, encountering another aircraft is common. For this reason, during times of low altimeter settings, the lowest flight levels become unusable, as they could conflict with aircraft operating at 17,000 feet. Thus, when altimeters are measuring 29.91 or lower, ATC cannot assign FL180 as it might not be separated with an aircraft at 17,000.

Airway Structure Improvements

As navigational systems have evolved, future aircraft will no longer be limited to the fixed airway system but will be able to navigate flexibly. To facilitate new, flexible routings, GNSS-based waypoints are being created around major airports and terminal areas. These new waypoints act as starting (pitch) and ending (catch) points to facilitate entry and exit from the terminal areas. Other than these predefined airspace fixes, pilots will be able to flexibly define their route of flight using a new method of defining routes known as the navigation reference system (NRS).

Navigation Reference System

The navigation reference system (NRS) is a grid of waypoints overlying the United States that will eventually become the basis for flight plan filing and air traffic operations. The NRS establishes waypoints every 30 minutes of latitude and every two degrees of longitude. Eventually NRS waypoints will have a grid resolution of 10 minutes of latitude by 1 degree of longitude. (See figure 3-12)

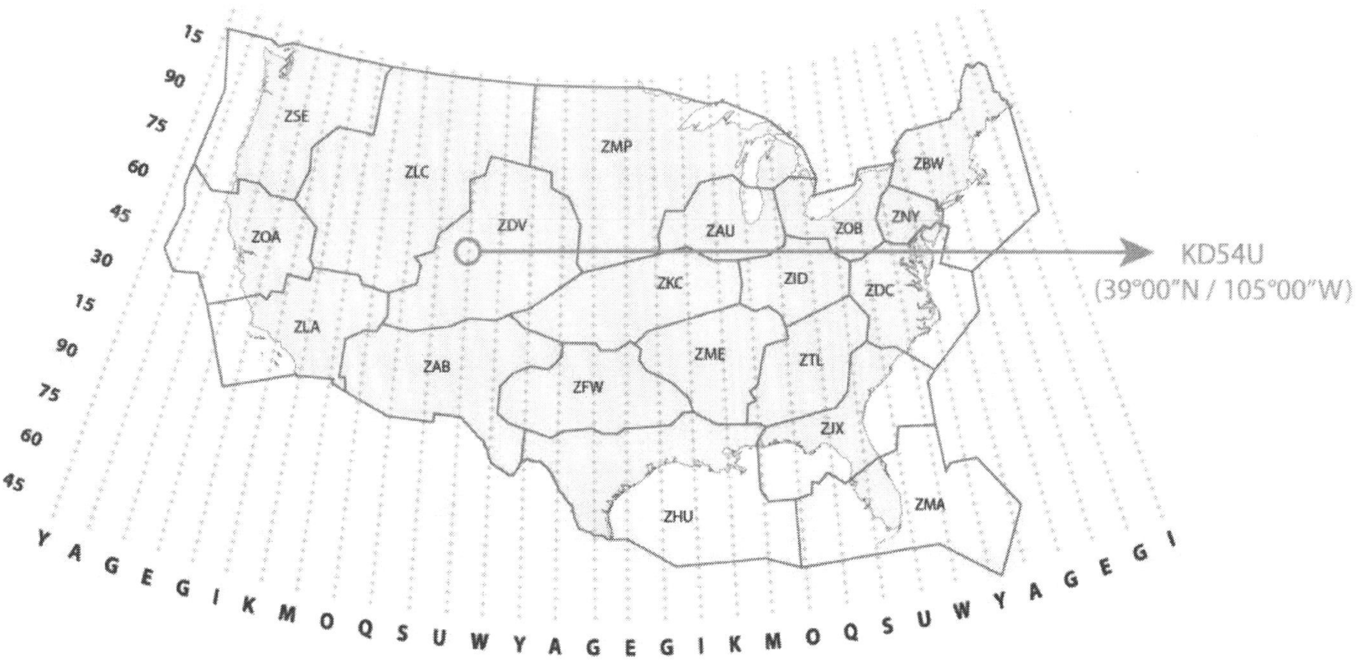

Figure 3-12. Navigation Reference System

Each NRS waypoint is assigned a five-character designator. The first character for all waypoints within the contiguous forty-eight U.S. states is a "K" (which is the ICAO identifier for the United States). The second character designates within which ARTCC the waypoint resides.

TABLE 3-5 ARTCC WAYPOINTS

Second Character	ARTCC
A	Albuquerque
B	Boston
C	Cleveland
D	Denver
F	Fort Worth
G	Chicago
H	Houston
I	Indianapolis
J	Jacksonville
K	Kansas City
L	Los Angeles
M	Memphis
N	New York
O	Oakland
P	Minneapolis
R	Miami
S	Seattle
T	Atlanta
U	Salt Lake City
W	Washington

A unique two number/one letter combination will follow which represents latitude and longitude but not encoded in a traditional lat-long format. The latitude increment numbers start at the equator, which is designated "00". Each 10-minute increment thereafter is then identified by a number between "01" and "90". The latitude numbering sequence repeats each 15 degrees of latitude. The longitude letters start at the prime meridian and go from west to east around the globe, repeating every 26 degrees. For example, the waypoint named KD54U can be identified as a U.S. waypoint (K), located in the Denver ARTCC's area (D) at latitude N39-00-00 (54) and longitude W105-00-00 (U).

Although this reference system may appear confusing, it will in fact be much easier to enter into a computer than a long string of latitude-longitude coordinates and is easier to build into navigation system internal error-checking protocols. Less overall navigation errors and miscommunications will result from the use of a standardized system such as NRS.

Special Use Airspace

In numerous areas scattered around the United States, it is in the national interest to either restrict or completely prohibit the flight of civilian aircraft. The U.S. government, through the Federal Aviation Regulations (FAR's), has designated these areas as special use airspace (SUA). SUA is designed to either confine unique aircraft operations or to entirely prohibit flight within the specified area. Unless otherwise noted, all of the following examples of special use airspace are published on VFR and IFR navigation charts and are described in appropriate aeronautical publications.

Prohibited Areas

A prohibited area is airspace where aircraft operations are absolutely prohibited. These areas are directly concerned with either national security or public safety. Designated prohibited areas include the White House, the Capitol Building, and Camp David. (See figure 3-13). The FARs expressly prohibit either IFR or VFR aircraft from entering such areas without specific (and very rarely granted) authorization. Air traffic controllers are not permitted to authorize civilian aircraft operations within these areas unless an emergency exists.

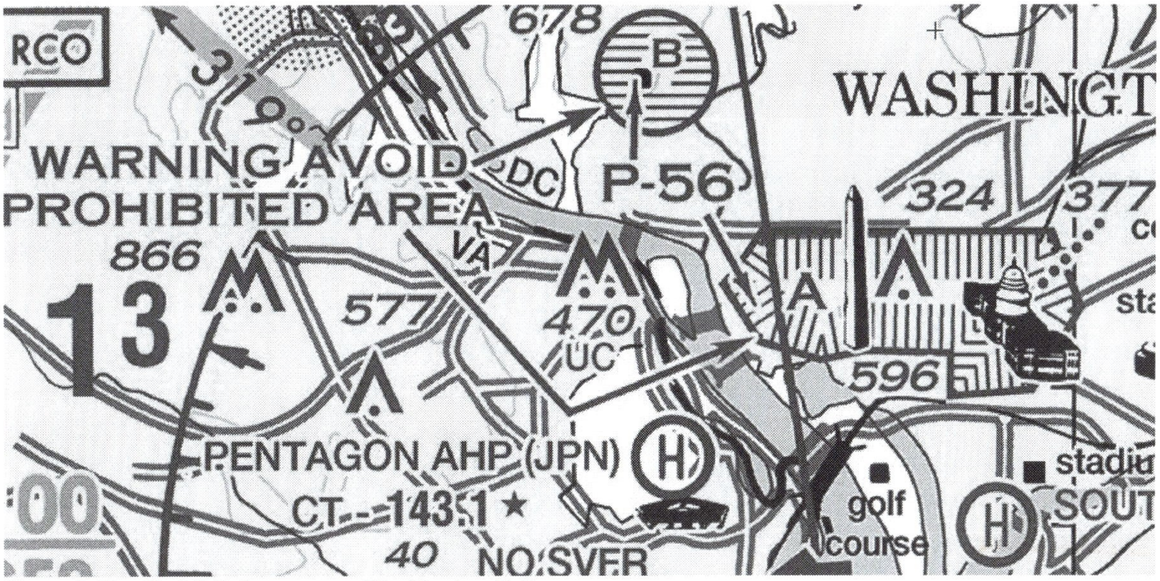

Figure 3-13. Prohibited areas over Washington, D.C.

Restricted Areas

Locations where aircraft operations are not prohibited at all times but are subject to some restriction are labeled restricted areas. (See figure 3-14). They are located where both airborne and ground-based activities are routinely conducted that may be hazardous to either the aircraft or its occupants. These activities include artillery firing, aerial gunnery, and high-energy laser and missile testing. Some restricted areas are in effect 24 hours a day; others operate part-time. Restricted areas are available for civilian flight when they are not active. ATC will reroute all IFR aircraft around "hot" restricted areas and can provide similar services to VFR pilots upon request. VFR aircraft not in contact with ATC are expected to know the status of the restricted area (distributed through Notices to Airmen, or NOTAMs) and avoid flying there.

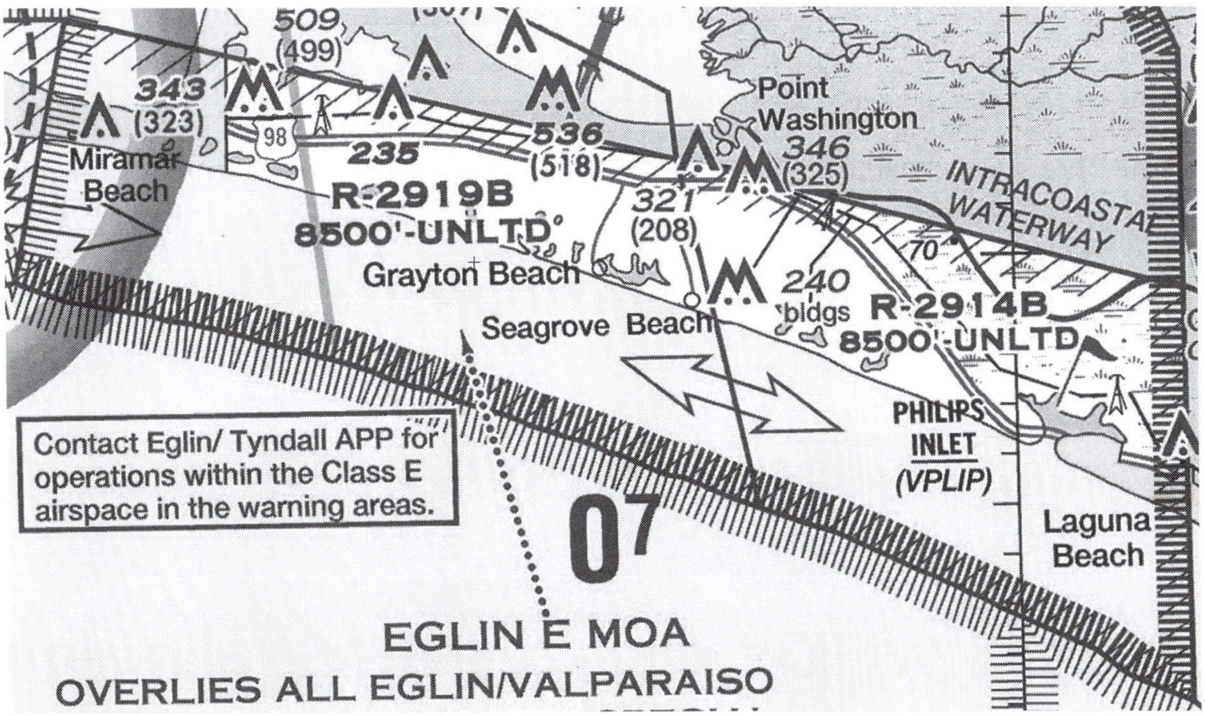

Figure 3-14. Restricted area in northwest Florida

Temporary Flight Restrictions

The FAA may impose temporary flight restrictions (TFRs) around any incident, accident, or event that has the potential for attracting a sufficient number of aircraft to create a hazard to either other aircraft in the air or people on the ground. Temporary flight restrictions may be imposed around earthquake, flood, fire, or aircraft crash sites for example. They are also used to protect the public interest for large scale public events, such as the Super Bowl and other events that attract large gatherings. TFRs essentially operate like temporary, ad-hoc restricted areas.

Whenever a temporary flight restriction is imposed, the FAA notifies pilots by issuing a NOTAM; these are also distributed nationwide to all ATCTs, ARTCCs, and FSSs as well, who can then relay the information to pilots. In addition, NOTAMs are transmitted to the airlines, military services, and many independent pilot-briefing companies who make the information available to their subscribers.

Special Flight Rules Areas

There have been special flight rules areas (SFRAs) around the United States for many years; they are primarily found in either areas with congested/complex airspace (Anchorage) or areas with large numbers of VFR flights (Grand Canyon and Los Angeles). Following the attacks of September 2001, the airspace around Washington D.C. underwent a number of changes designed to restrict flight operations. After a number of VFR accidents in and around New York City, special flight rules were set up there as well. All of these areas of special flight rules are codified in part 93 (special air traffic rules) of the Federal Aviation Regulations and affect the following areas:

TABLE 3-6 Special Fight Rules Areas

City	State
Anchorage	AK
Ketchikan	AK
Grand Canyon	AZ
Luke AFB	AZ
Los Angeles	CA
Washington	D.C.
Valparaiso	FL (Eglin AFB)
New York	NY
Niagara Falls	NY
Lorain County	OH (Cleveland area)

In order to fly within any of these SFRAs, pilots must conform to the routes, procedures and altitudes specified in FAR 93 or may be subject to penalty by the FAA. Air traffic procedures for these areas may be modified from the standard to incorporate these requirements.

DC Flight Restricted Zone and Special Flight Rules Area

The FAA established the Washington, D.C. Flight Restricted Zone (FRZ) and SFRA in 2003 to restrict air traffic around the nation's capital. There are very specific rules and penalties for violating the rules pursuant to flying within this airspace. Pilots who do not adhere to the proper procedures could be intercepted in flight, directed to a safe landing area, and detained and interviewed by law enforcement personnel. They also face fines and loss of certifications.

The DC FRZ extends outward roughly 30 nm in radius from Washington, D.C. and extends vertically from the ground up to, but not including, FL 180. Aircraft must remain clear of this area unless they are properly equipped, have filed a special flight plan, and have received clearance to enter or exit the area.

Warning Area

A warning area is airspace located over international waters where operations that may be hazardous to nonparticipating aircraft are routinely conducted. (See figure 3-15). The activities conducted in a warning area are usually similar to those performed in a restricted area. Since warning areas are located in international airspace, neither the United States nor any other government has the right to restrict the flight of aircraft through these areas. Both IFR and VFR aircraft may operate in warning areas, but they do so at their own risk. ATC will normally reroute all IFR aircraft around "hot" warning areas and can provide similar services to VFR pilots upon request.

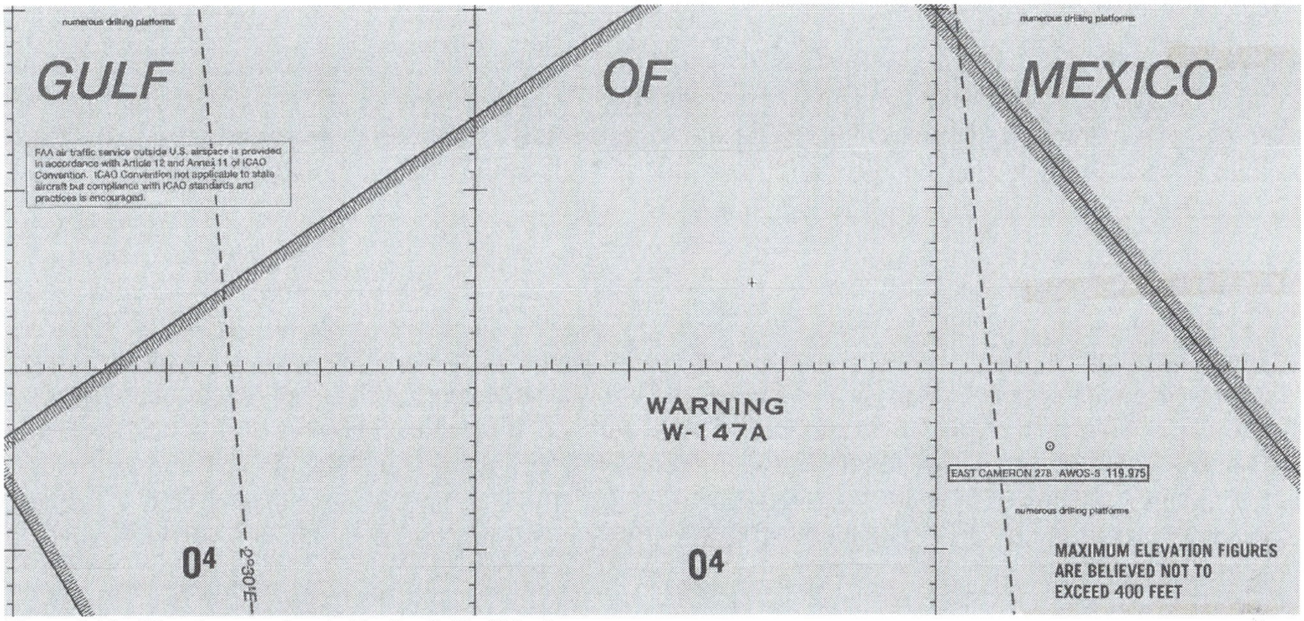

Figure 3-15. Warning area located in the Gulf of Mexico.

Military Operations Area

A military operations area (MOA) is designated low-level airspace (below FL180) where military flight training activities routinely take place that could prove hazardous to civilian aircraft. Some flight training conducted by military aircraft requires acrobatic maneuvers to be practiced on or near a federal airway. Although acrobatic flight along a federal airway is normally forbidden by federal regulation, the Department of Defense has been exempted if the maneuvers are conducted within an MOA. (See figure 3-16). IFR aircraft will typically be routed around an active MOA by ATC; VFR pilots are permitted to enter an MOA but do so at their own risk. As VFR aircraft are not allowed above FL180, that is not a concern for an ATCAA.

Military Training Routes

Military pilots are often required to practice low-level, high-speed, combat-training flights. The maneuvers performed during these training flights make the "see and avoid" concept of traffic separation difficult without increased vigilance on the part of both military and civilian pilots. To assist civilian pilots to avoid these military aircraft, the FAA and the DoD have mutually agreed to participate in the military training route (MTR) program. Through this program, designated MTR routes have been designated by both the FAA and the DOD and are depicted on VFR navigation charts as VFR (VR) or IFR (IR) routes. ATC will normally

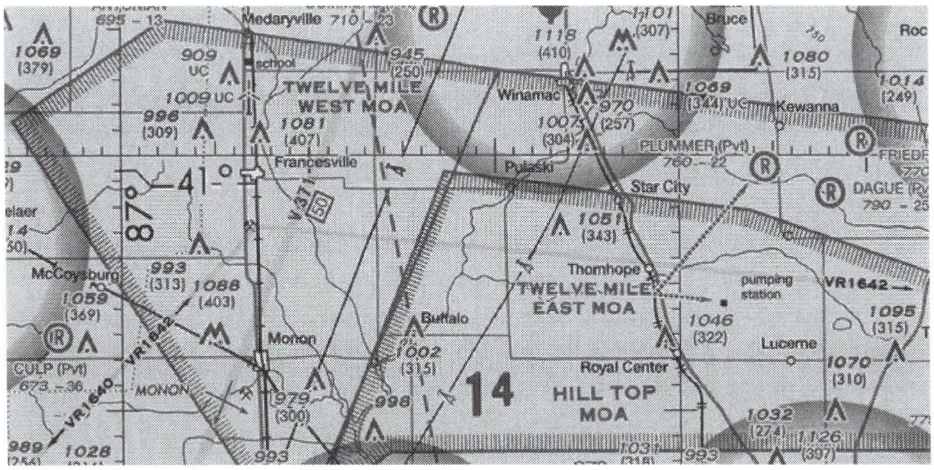

Figure 3-16. MOAs and military training routes

NOT reroute IFR aircraft around MTRs as aircraft on an MTR are normally operating below the altitudes reserved for airways and IFR flights. The FAA often does separate military aircraft entering or exiting an MTR however as the aircraft transit more congested civilian airspace. (See figure 3-16).

Alert Areas

Alert areas are airspace that may contain a large number of high-performance military or civilian aircraft conducting routine training exercises, or any unusual aerial activity. Although there are no legal restrictions to civilian aircraft flying through an alert area, both IFR and VFR pilots transiting the area should be aware of the large numbers of VFR aircraft practicing maneuvers there.

Controlled Firing Areas

Controlled firing areas contain activities that, if not conducted in a controlled environment, could be hazardous to aircraft. These areas are not identified on VFR and IFR charts since the controlling agency suspends its activities whenever nonparticipating aircraft approach the area. Such aircraft are usually detected by the use of spotter aircraft, radar, or ground-based observers. Whenever intrusion of a nonparticipating aircraft into a controlled firing area is detected, the test firings are halted until the aircraft in question has departed the area. Controlled firing areas predominantly affect low flying aircraft since most test firing is conducted at these altitudes. ATC can reroute IFR aircraft around controlled firing areas when requested.

National Security Areas

National security areas (NSAs) consist of airspace established at locations where increased security and safety of ground facilities are required. Some such areas might include nuclear power plants, or large dams. Although flights through such areas are not prohibited, pilots are requested to voluntarily avoid flying through NSAs whenever possible. Controllers are supposed to be aware of any unusual or increased activity in these areas, and to report it to proper authorities. ATC will not reroute IFR or VFR aircraft around national security areas.

Communications Systems

Radio Phraseology

The safe operation of the nation's air traffic control system ultimately depends on reliable and accurate communication between pilots and air traffic controllers. Virtually every instruction, procedure, or clearance used to separate aircraft relies on written or verbal communication. Any miscommunication between participants in the air traffic control system might contribute to, or even be the direct cause of, an aircraft accident with a subsequent loss of life. For this reason, proper communications procedures must always be observed by both pilots and controllers.

Many accidents and incidents that have occurred over the last fifty years can be attributed to improper or misunderstood communications. Although many improvements to the air traffic control communications system have made it less reliant on verbal or written communication, pilots and controllers will continue to rely on human communication well into the future. Although both controllers and pilots must possess a proper understanding of communications procedures and phraseology, it is crucial for controllers to use proper phraseology at all times. Additionally, as many errors are caused by failing to correct a pilot's improper understanding of a clearance (known as a hearback error), it is essential for controllers to practice conscious listening skills.

U.S. pilots and controllers are fortunate that ICAO has designated English as the international language for ATC communications worldwide, as they have no need to learn a foreign language for their job. However, air traffic controllers should realize that although foreign pilots are able to communicate using English, they potentially do not have full command of the language. Thus, phraseology and slang not approved by ICAO or the FAA should be avoided when communicating with all pilots, both foreign and English-speaking ones. Doing so will help reduce the risk of miscommunication; additionally, it will help non-English speaking pilots maintain situational awareness, which is another leading cause of ATC errors.

Radio Frequency Assignments

International agreements allocate certain radio frequency bands for use in aeronautical communications. These frequency bands exist primarily in the high (HF), very high (VHF), and ultra-high (UHF) spectrums. As HF are not line of sight and can thus follow the curvature of the earth, they are primarily used for long-range communication. They are particularly difficult to listen to however, due to the high-pitched, static-filled nature of the frequency. Only a few ATC facilities, such as ARTCCs with oceanic responsibility find a need to use these frequencies. Most HF communications are run through an outside group, generally known as ARINC, funded by a consortium of airlines to aid in remote communications.

Most U.S. ATC facilities use both VHF and UHF for routine air-to-ground communication. UHF radio equipment is primarily used by military aircraft, whereas VHF is used by both military and civilian aircraft. The frequencies used in ATC communications are assigned by the Federal Communications Commission (FCC) in cooperation with the FAA. Since there is not a sufficient number of available frequencies in either the VHF or UHF spectrum to permit every ATC facility to operate using a separate frequency, the FCC often assigns the same frequency to two or more ATC facilities. Because the radio transmissions from high-altitude aircraft travel farther than those from low-flying aircraft, the FCC must carefully determine any potential interference problems before assigning these frequencies.

Most air traffic controllers still use voice radio to communicate with aircraft. In general, each controller is assigned one or more radio frequencies for communications with pilots and has access to telephone equipment for communication with other controllers in the same facility or in adjacent facilities. To facilitate seamless communication, each ATC position is equipped with a voice control switching system (VSCS), which allows communication with pilots and other controllers. Most controllers are outfitted with a boom mike and headset assembly which permits them to remain in contact with both pilots and other controllers.

Standard Phraseology

To ensure that miscommunication is kept to a minimum, it is imperative that controllers use standard phraseology and procedures recommended by ICAO and the FAA. When communicating with pilots or other controllers the following message format should always be used.

1. Identification of the aircraft or controller being contacted. This alerts the intended receiver of the upcoming transmission. This is used when contacting pilots and other controllers.
2. Identification of the facility/controller initiating the call. This identifies who is initiating the communication. This can be omitted upon subsequent conversations with pilots to abbreviate frequency congestion.
3. The contents of the message. The message format should conform to standards approved by the FAA.
4. Termination. In communications with another ATC facility, the message is always terminated with the controller's assigned operating initials. This simplifies identification of the controller if a subsequent investigation is necessary, as well as indicates to the receiving controller that communication is finished. In communicating with pilodts, this is frequently omitted, as it is generally obvious.

As mentioned, the above format should be used when contacting pilots, as well as other controllers. However, when using the VSCS to call a different control position, the controller answering the VSCS line should observe proper protocols. When answering any phone line, a controller should always identify their position, for example, “Denver center, Sector 32”, “Portland approach”, or “Boston ground”. By doing so, they ensure that the controller placing the call knows they have reached the correct position. Additionally, doing so also relieves the calling controller from having to identify who they contacted (see Step 1 above).

Standard Pronunciation

Certain letters and numbers may sound similar to one another when spoken over low-fidelity radio or telephone equipment. In addition, accents and dialects may make it difficult to discern and identify the exact content of a message. To alleviate this problem, a standard for pronunciation of letters and numbers has been approved by ICAO and adopted by the FAA.

Numbers

Numbers can be stated in two different ways-- group form or individually. In general, unless specified otherwise, each number should be enunciated individually unless group form pronunciation is stipulated. For example, the number 10 would be pronounced individually as “one zero”, but in group form would be “ten”. The number 1246 would be pronounced individually as “one two four six”, but in group form as “twelve forty six”.

When using group form pronunciation, the following format should be used:

TABLE 3-7 GROUP FORM PRONUNCIATION

Number	Pronunciation
2	Two
45	Forty-five
336	Three thirty-six
7321	Seventy-three twenty-one

The following two tables contain official ICAO pronunciation for other numbers and letters.

TABLE 3-8 STANDARD PRONUNCIATION OF NUMBERS

Character	Number	Pronunciation
0	Zero	Zee-ro
1	One	Wun
2	Two	Too
3	Three	Tree
4	Four	Fow-er
5	Five	Fife
6	Six	Six
7	Seven	Sev-en
8	Eight	Ait
9	Nine	Nin-er

TABLE 3-9 STANDARD PRONUNCIATION OF NUMBERS

Character	Word	Pronunciation
A	Alpha	Al-fah
B	Bravo	Brah-voh
C	Charlie	Char-lee
D	Delta	Del-ta

Character	Word	Pronunciation
E	Echo	Eck-oh
F	Foxtrot	Foks-trot
G	Golf	Golf
H	Hotel	Hoh-tell
I	India	In-dee-ah
J	Juliett	Jewlee-ett
K	Kilo	Key-loh
L	Lima	Lee-mah
M	Michael	Michael
N	November	Nov-em-ber
O	Oscar	Oss-cah
P	Papa	Pah-pah
Q	Quebec	Key-beck
R	Romeo	Row-me-oh
S	Sierra	See-air-ah
T	Tango	Tang-go
U	Uniform	You-nee-form
V	Victor	Vik-tah
W	Whiskey	Wiss-key
X	X-ray	Ecks-ray
Y	Yankee	Yang-key
Z	Zulu	Zoo-loo

Altitudes

Unless otherwise specified, every altitude used in the ATC system is measured above MSL. The only routine exception is cloud ceilings, which are measured above ground level (AGL). A controller who must issue an AGL altitude to a pilot should advise the pilot that the altitude being stated is above ground level. Altitudes should be separated into thousands and hundreds; the thousands are always pronounced separate from the hundreds. Each digit of the thousands number should be enunciated individually, whereas the hundreds should be pronounced in group form. For example, the altitude 3,900' would be pronounced "three thousand niner hundred". An altitude of 12,500' would be pronounced as "one two thousand fife hundred". 17,000' would be pronounced "one seven thousand".

Flight Levels

At or above 18,000' MSL, all altitudes are referred to as flight levels (FL). As discussed previously, all aircraft operating above this altitude used the same altimeter setting of 29.92. When communicating, flight levels should be preceded by the words "flight level," and each number should be enunciated individually. For example, 18,000' MSL would be pronounced "flight level one eight zero"; FL390 would be pronounced "flight level three niner zero".

TABLE 3-10 FAA-APPROVED ALTITUDE PRONUNCIATION

Altitude	Pronunciation
50	Fife thousand
110	One one thousand
135	One three thousand five hundred
172	One seven thousand two hundred
220	Flight level two two zero
315	Flight level tree one fife

Time

A common system of time measurement is essential to the safe operation of the ATC system. The FAA and ICAO have agreed that local time is not to be used within ATC. Instead, every ATC facility around the world must use the same time standard, known as coordinated universal time (UTC). UTC is the same as local time in Greenwich, England, which is located on the 0° line of longitude, known as the prime meridian. UTC was previously known as Greenwich Mean Time (GMT).

The use of UTC around the world eliminates the question of which time zone a facility or aircraft is located in. In addition, the use of UTC eliminates the need for "a.m." and "p.m." by using a 24-hour clock system. UTC is always issued as a four-digit number, and the word "o'clock" is never pronounced. For example, 6:20 a.m. becomes 0620, and 6:20 p.m. becomes 1820. Local time is converted to UTC by adding the number of hours indicated in table 3-8.

To convert from local time to UTC, convert the local time to a 24-hour clock, and then add the required time difference. To convert from UTC to local time, subtract the difference and convert from a 24-hour to a 12-hour format. For example:

4:35 a.m. (EST) is 0435 (EST), which is 0935 (UTC)
9:13 p.m. (PDT) is 2113 (PDT), which is 0413 (UTC) the next day
1125 (UTC) is 0425 (MST), which is 4:25 a.m. (MST)

TABLE 3-11 STANDARD TIME ZONES

Time Zone	Difference
Eastern standard time (EST)	5 hours
Eastern daylight time (EDT)	4 hours
Central standard time (CST)	6 hours
Central daylight time (CDT)	5 hours
Mountain standard time (MST)	7 hours
Mountain daylight time (MDT)	6 hours
Pacific standard time (PST)	8 hours
Pacific daylight time (PDT)	7 hours
Alaskan standard time (AST)	9 hours
Alaskan daylight time (ADT)	8 hours

As most aircraft and all ATC facilities have UTC time displayed, pilots and controllers do not frequently have to perform this conversion. If there is a possibility of confusion; the controller can suffix any UTC time with the word "zulu". Since local time is so infrequently used, both pilots and controllers should clarify by suffixing any use of it with the word "local."

Altimeter Settings

For aircraft flying below 18,000', the pilot must be issued the proper barometric pressure so that the aircraft's altimeter can be properly adjusted to indicate the correct altitude above mean sea level. Aircraft transitioning from flight levels to lower altitudes must also be issued a local altimeter setting. The controller should issue these altimeter settings by individually enunciating every digit without pronouncing the decimal point; the altimeter setting should be preceded by the word "altimeter":

TABLE 3-12 ALTIMETER SETTING PRONUNCIATIONS

Altimeter Setting	Pronunciation
29.92	Altimeter two niner niner two
30.16	Altimeter tree zero one six

Wind Direction and Velocity

Wind direction at airports is always determined in reference to magnetic north and indicates the direction from which the wind is blowing. The direction is always rounded off to the nearest 10°. Thus, a wind blowing from north to south is a 360° wind; a wind from the east is a 90° wind. The international standard for measuring wind velocity requires that wind speeds be measured in knots; 1 knot equals approximately 1.15 miles per hour. Wind direction and velocity information is always preceded by the word "wind," with each digit of the wind direction enunciated individually. The wind direction is then followed by the word "at" and the wind velocity in knots, with each digit enunciated individually. If the wind direction is constantly changing, the word "variable" is used. If the wind velocity is constantly changing, the word "gusts" and the peak speed are suffixed to the wind speed. Here are some examples:

TABLE 3-13 WIND DIRECTION PRONUNCIATIONS

Wind Direction	Wind Speed	Pronunciation
From the north	15 knots	Wind three six zero at one five
From the east	10 knots with occasional gusts to 25 knots	Wind zero niner zero at one zero gusts two five

Headings

Aircraft headings are also measured in reference to magnetic north. If the heading contains fewer than three digits, it should be preceded by a sufficient number of zeros to make a three-digit number. Aircraft headings should always be preceded by the word "heading," with each of the three digits enunciated individually. Unless absolutely necessary, headings are normally issued in even ten-degree increments. Here are some examples:

TABLE 3-14 EXAMPLE HEADING PRONUNCIATIONS

Heading	Pronunciation
090°	Heading zero niner zero
125°	Heading one two fife
250°	Heading two fife zero

Radio Frequencies

When pronouncing issuing radio frequencies, the controller should enunciate each digit individually. Current VHF communications radios use 25 kHz spacing between assigned frequencies. For instance, the next usable frequency above 119.600 is 119.625, followed by 119.650, 119.675, then 119.700. The first number after the decimal is always pronounced. But if the second number after the decimal is a zero, it is not pronounced. The third number after the decimal is never pronounced, since it is always either a zero or a five and can be assumed. VHF and UHF communication and navigation frequencies always use the decimal point. The decimal is pronounced as "point". Enunciating the decimal point reduces potential frequency confusion distinguishing between a frequency and altimeter setting; thus, it is one of the ways for both pilots and controllers to help reduce communication errors. ICAO standards differ a little from FAA in this regard. An ICAO trained pilot or controller will use the word "decimal" where and FAA controller/US trained pilot will say "point".

TABLE 3-15 FREQUENCY PRONUNCIATIONS

Frequency	Pronunciation
119.600 MHz	One one niner point six
343.000 MHz	Three four three point zero
123.050 MHz	One two three point zero five
131.725 MHz	One three one point seven two five

Aircraft Speeds

Aircraft speeds, like wind speeds, are always measured in knots. Airspeeds are always expressed with each digit enunciated individually and suffixed with the word "knots". Aircraft above (roughly) 24,000 feet routinely measure their speed in Mach number, which is the speed of sound. When discussing Mach speed, the word point is annunciated; a zero before the point is implied. When discussing speed in ATC, controllers and pilots are using indicated airspeed, which can differ significantly from their speed over the ground. This will be discussed more in the enroute control chapter. Here are some examples:

TABLE 3-16 SPEED

Speed	Pronunciation
250	Two fife zero knots
95	Niner fife knots

Air Traffic Control Facilities

ATC facilities are identified by name, using the name of the city where the facility is located followed by the type of facility or the operating position being communicated with. If a particular city has two or more airports, the airport name is used instead of the city name. Approach controls and centers are always named after the largest nearby city. When communicating with other controllers, ATC will add information that specifies their individual position, as that will matter to other controllers, but not generally to pilots. Some examples follow:

TABLE 3-17 AIR TRAFFIC CONTROL FACILITIES

Facility Type	Pronunciation
Local control	Tower
Ground control	Ground
Clearance delivery	Clearance
Air route traffic control center	Center
Flight service station	Radio
Approach control	Approach
Departure control	Departure

Airways, Routes and Navigation Aid Descriptions

Airways are always described with the route identification pronounced in group form. The route number is prefixed with "victor" if it is a low-altitude airway, "jay" if it is a high-altitude airway, "tango" or "Q" if it a GPS based route. The following chart has some examples:

TABLE 3-18 ROUTE

Route	Pronunciation
V251	Victor two fifty-one
J97	Jay ninety-seven
T368	Tango tree sixty-eight
Q12	Cue twelve

Aircraft Identification

Aircraft are identified using procedures that help eliminate confusion and misdirected instructions. The assigned aircraft identification call signs used by pilots and controllers vary depending on the type of operation in which the aircraft is involved. If the aircraft is a scheduled airline flight, the FAA has authorized the use of a distinctive airline name that should always be used when communicating with that aircraft; this is known as a three-letter identifier, or FAA-approved call sign. In addition to this name, every airline flight has been issued a flight number by the airline itself. The approved aircraft identification consists of the airline name, followed by the flight number pronounced in the group format discussed earlier (UAL1276 = "United twelve seventy-six").

Most authorized airline names are easily recognizable, although a few are somewhat unusual. These approved airline names have been selected to ensure that no two sound similar. Every airline has also been issued a three-letter designator to be used in written communications concerning the aircraft. A list of air carrier names and their three-letter identifiers can be found in the Contractions Handbook published by the FAA. Here are some examples from the handbook.

TABLE 3-19 AIRLINE NAME WITH FAA IDENTIFIER AND CALL SIGN

Airline Name	FAA Identifier	Call Sign
Aer Lingus	EIN	Shamrock
Air Canada	ACA	Air Canada
Air China	CCA	Air China
Air France	AFR	Airfrans
Air Wisconsin	AWI	Air Wisconsin
Alaska	ASA	Alaska
American	AAL	American
British Airways	BAW	Speedbird
Cathay Pacific	CPA	Cathay
Delta	DAL	Delta
Emirates Airlines	UAE	Emirates
Envoy Air	ENY	Envoy
Federal Express	FDX	Fedex
Frontier	FFT	Frontier Flight
Japan Air Lines	JAL	Japanair
JetBlue	JBU	Jetblue
KLM	KLM	KLM
Mesa	ASH	Air Shuttle
Piedmont	PDT	Piedmont
Republic	RPA	Brickyard
Shuttle America	TCF	Mercury
Sky West Airlines	SKY	Skywest

Airline Name	FAA Identifier	Call Sign
Southwest	SWA	Southwest
Spirit Airlines	NKS	Spiritwings
Trans States Airlines	LOF	Waterski
United Airlines	UAL	United
United Parcel	UPS	UPS
Virgin America	VRD	Redwood
Virgin Atlantic	VIR	Virgin
WestJet	WJA	Westjet

General aviation aircraft call signs consist of the type of aircraft plus a unique serial number assigned by the FAA. The call sign may contain up to five numbers or letters. When the call sign is pronounced, each character is enunciated individually. Every U.S. aircraft's serial number is preceded by the letter N, signifying that it is registered in the United States. During routine communications, this letter is usually not pronounced but can be used if the pilot wishes. Aircraft registered in other countries have aircraft identification numbers or letters preceded with a letter other than N, for example, all Canadian-registered general aviation aircraft start with the letter C. Additionally, general aviation aircraft frequently include the model or company of manufacturing in their call sign. For example, a Piper Navajo (PA35) with the tail number 38T could be referred to as November 3-8 Tango, Navajo 3-8-Tango, or Piper 3-8-Tango. The following table illustrates that concept:

TABLE 3-20 SAMPLE AIRCRAFT CALL SIGNS

Aircraft/Tail number	November call sign	Manufacturer call sign	Model call sign
PA35/46M	November four six Mike	Piper four six Mike	Navajo four six Mike
C206/15E	November one fife echo	Cessna one fife echo	Caravan one fife echo
BE20/ 62A	November six two alpha	Beech six two alpha	KingAir six two alpha

Military aircraft are assigned a variety of call signs that may include five numbers, one word followed by numbers, or two letters followed by numbers. The aircraft's call sign is always prefixed with the name of the military service. Each word is pronounced in full with the letters and numbers enunciated individually, as illustrated in the following table:

TABLE 3-21 MILITARY AIRCRAFT

Call Sign	Military Service	Pronunciation
C693	Coast Guard	Coast Guard six niner tree
R23956	Army	Army two three niner fife six
VV1963	Navy	Navy one niner six tree
VM4257	Marine	Marine four two fife seven
A14932	Air Force	Air Force one four niner tree two

Surveillance

Radar

Air traffic controllers use radar, which is similar to broadcast radio, to detect and track aircraft. (See figure 3-17). An acronym for Radio Detection and Ranging, radar is not a new development but an improvement of concepts that date back to the late nineteenth century. Modern radar systems are composed of the following four components: the transmitter, antenna, receiver, and display.

The transmitter creates the high-powered radio pulses used by the radar system. Modern radar transmitters operate on frequencies in the UHF band or higher. The radar pulse lasts about one microsecond; the pulses are transmitted rapidly at a rate of about 1,000 pulses per second. The pulse can have over a million watts of electrical power.

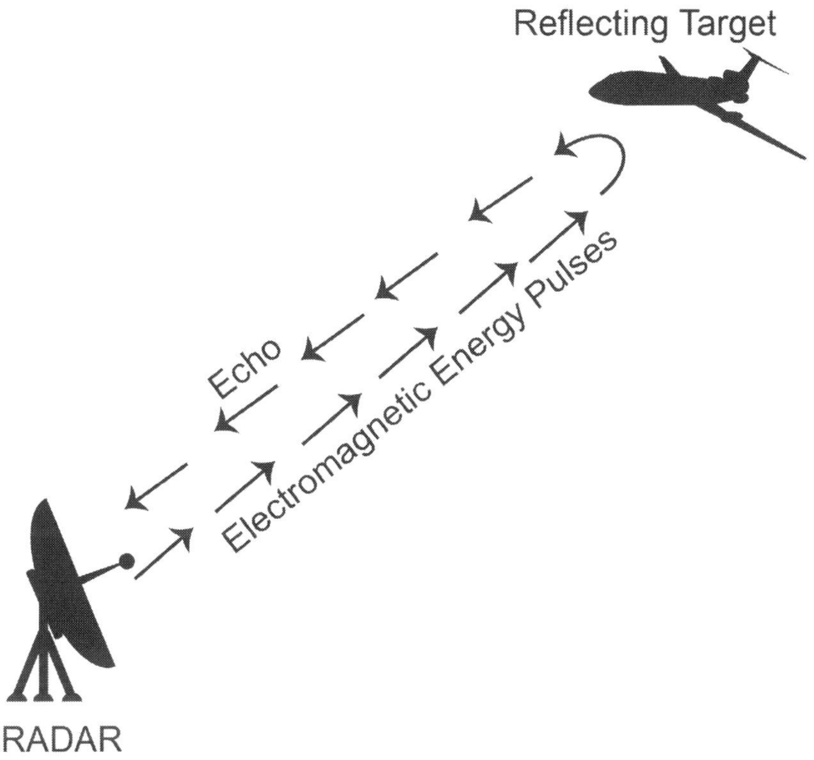

Figure 3-17. Radar operation

Antenna

The radio pulses emitted by the transmitter are routed to the antenna using a waveguide—a hollow metal channel that conducts the microwave energy to the antenna. The antenna, which is parabolic in shape, is mounted on a rotor. Most current radar antennas are not solid; they are constructed of a metallic mesh or grid. (See figure 3-18). Sometimes the radar antenna is referred to as the "sail".

Although the mesh looks porous, properly constructed antennas appear solid to microwave transmissions. The antenna's only function is to provide a reflecting and focusing surface for the radar pulse. The waveguide terminates at the feedhorn, located at the focal point of the antenna. The radar pulse is routed through the waveguide from the transmitter and emanates from the feedhorn. After leaving the feedhorn, the pulse is reflected and focused by the antenna into a narrow, vertical beam approximately 2° wide and 40° high, known as the antenna boresight. If the radar pulse hits an object, it reflects off the object and part of the reflection returns to the transmitter. This is known as target illumination, and the reflection is known as the echo.

Figure 3-18. Radar antenna, waveguide and feedhorn assembly

Receiver

Immediately after the transmitter shuts off, the receiver is switched into the circuit to listen for any echoes. If the radar pulse is reflected by an object, a small portion of the radio energy returns to the antenna and will be focused back into the feedhorn. This pulse then returns to the receiver through the waveguide. Since radio energy diminishes proportionately to the distance it must travel, by the time the echo has returned it will have lost a considerable amount of power. This pulse must then be amplified, sometimes at least a millionfold, so that it can be processed and displayed by the receiver.

The radar system measures the time difference between transmission and reception and uses this calculation to determine the object's distance from the antenna. Since radio signals traveling at the speed of light take 6.18 microseconds to travel 1 nautical mile, these pulses will take 12.36 microseconds—6.18 doubled —to travel one mile, reflect off an object, and return. This time—12.36 microseconds—is known as a radar mile.

Radar Display

Once the radar system has received and processed the reflected signal, the object's relative position can be displayed to the controller in many ways. On older systems, targets are often displayed on a cathode ray tube known as a plan position indicator (PPI), radar scope, or just scope. Newer systems use high resolution computer monitors specially designed for air traffic control.

Video Map

Every radar system is equipped with a video map selector and a video map intensity control. To use radar to separate and assist aircraft, the controller must know the position of each aircraft in relation to airports, navigation fixes, and airways. Many different maps may be available to each controller. The maps can include symbology to indicate particular navaid positions, obstructions, airways, and intersection locations, as well as other items deemed necessary for air traffic facilities. Airport locations and the extended runway centerlines used for instrument approaches are also included on most video maps. Sector boundaries, SUA boundaries, and other important information defined in facility directives or letters of agreement are also displayed. (See figure 3-19).

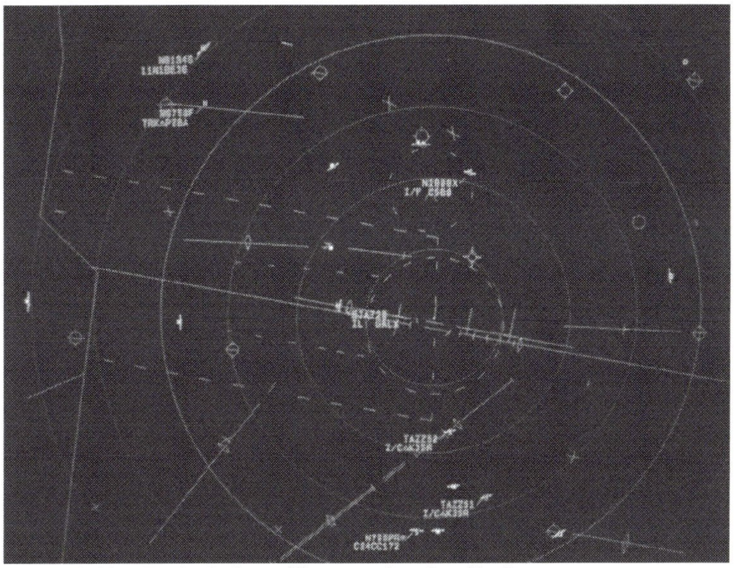

Figure 3-19. Typical air traffic control radar display

Types of Air Traffic Control Radar

In general, three radar systems are used in air traffic control in the United States:

- airport surveillance radar (ASR),
- air route surveillance radar (ARSR),
- and airport surface detection equipment (ASDE).

ASR is a short-range radar that approach and departure controllers primarily use; ARSR is a long-range radar that ARTCC controllers use to provide enroute separation of aircraft. Control towers use ASDE, a short-range radar system, during periods of extremely low visibility to detect aircraft or vehicles moving around the airport. Some ADSE is certified for aircraft separation; some is used only for situational awareness.

Airport Surveillance Radar

The primary short-range radar currently used by the FAA is airport surveillance radar. Most major civilian and military airports have ASR systems. TRACON controllers use ASR, which has a range of approximately 100 nautical miles, to separate local aircraft with the TRACON airspace. The first ASR used by the FAA was the ASR-1 series. In 2017, the FAA was using ASR-8, ASR-9, and ASR-11 series radars. ASR-9 radar was the FAA's first solid-state equipment radar. ASR-11 is the first all-digital radar for use by air traffic controllers. (See figure 3-20).

Figure 3-20. Airport surveillance radar model ASR-11

Air Route Surveillance Radar

The first ARSR ordered by the FAA was the ARSR-1 series purchased in the late 1950s. Currently, the FAA uses modified ARSR-1E, ARSR-2, ARSR-3, and ARSR-4 radar systems. (See figure 3-21). These radar systems will provide enroute radar coverage at least until the year 2025. ARSRs can be located hundreds of miles from the ARTCC and their signals can be digitally transmitted to multiple facilities.

Figure 3-21. Air route surveillance radar

The U.S. Air Force also operates air defense radars that are similar to ARSR. The Air Force and the FAA jointly operate a number of long-range radar systems, known as joint surveillance systems (JSSs). Joint surveillance systems use the same transmitter, antenna, and receiver for the Air Force and the FAA, but they are equipped with an electronic splitter that sends duplicate radar information to military and civilian air traffic control facilities for processing. The FAA uses the information for air traffic control; the Air Force concentrates on air defense. Maintenance and operation costs are shared jointly by the FAA and the DoD.

Airport Surface Detection Equipment

At airports where reduced surface visibility often makes it impossible to see each aircraft, a specialized radar system designed to locate and display ground-based aircraft and vehicles has been designed. ASDE provides surveillance of aircraft and airport service vehicles. (See figure 3-22). At high-activity airports, radar monitoring of aircraft and ground vehicles is essential for safety during periods of reduced visibility. Aircraft and vehicle position information are reported even during periods of heavy snow and fog. The ASDE system is used whenever weather conditions preclude visual observation of the runways and taxiways. Because of its necessary sensitivity, a typical ASDE display is cluttered and difficult to interpret. Installation of the ASDE system is cost effective only at high-activity airports where reduced visibility is commonplace.

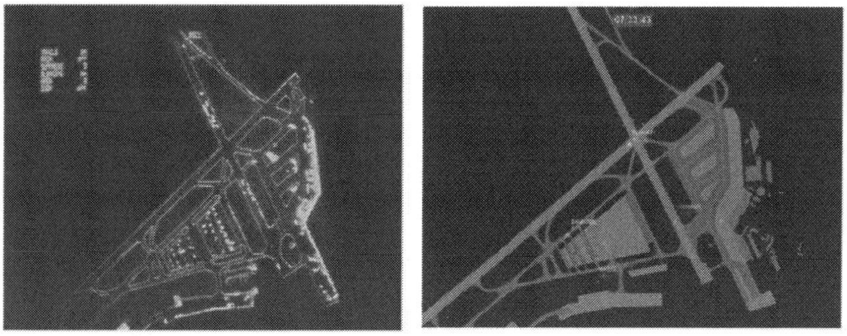

Figure 3-22. ASDE display (on left), with newer ASDE-X display on the right.

Air Traffic Control Radar Beacon System

One of the most significant developments in air traffic control technology has been the development of a secondary radar system known as the air traffic control radar beacon system (ATCRBS), secondary surveillance radar (SSR), or simply secondary radar.

Primary radar efficiency and operation depend on a number of variables, including transmitter power, aircraft size and distance from the radar antenna, atmospheric conditions, and obstructions that may interfere with the transmitted radar signal. In addition, primary radar systems detect and display every aircraft within the range of the radar antenna, even if it is above or below the vertical limits of the controller's assigned sector. When using primary radar for air traffic control separation, a controller cannot positively determine each aircraft's altitude; instead he or she must depend on accurate altitude reports from the pilot—a time-consuming, inefficient, and potentially inaccurate means of verifying each aircraft's altitude.

ATCRBS uses two ground-based antennas to transmit a challenge signal to every aircraft. Every aircraft equipped with a transponder is set to respond with a coded reply. (See figure 3-23). Ground-based computers use this information to determine the aircraft's identity, altitude, predict flight paths, and provide other essential information to the controller. In its simplest method of operation, the ATCRBS causes a distinct slash, known as a beacon slash, to be created on the PPI display. This beacon slash directly over-lays the primary target blip and identifies the target as a transponder-equipped aircraft, not a vehicle, train, or some other object.

Code Selection

The secondary radar system depends on equipment on the aircraft, the transmitter, transmitting a certain series of number assigned to the aircraft, which is called their squawk. When the pilot selects a particular code on the transponder, it is known as squawking. When the interrogator on the ground transmits a challenge on Mode C, any aircraft equipped with an altitude-encoding transponder also replies with the aircraft's altitude. Thus, all Mode C enabled aircraft transmit their current altitude to ATC facilities, making them easier for controllers to separate. All of this information can be converted to digital information and processed by a computer for display to the controller.

Figure 3-23. Transponder "squawking" code 1200

Computerized Radar Systems

The primary and secondary radar information is usually processed with modern digital computers to provide enhanced ATC displays to controllers. There are many different computer systems currently in use by the FAA. (See figure 3-24). Although the ERAM, STARS and ARTS systems are todays common system, that will likely to change in the future as automation capability improves.

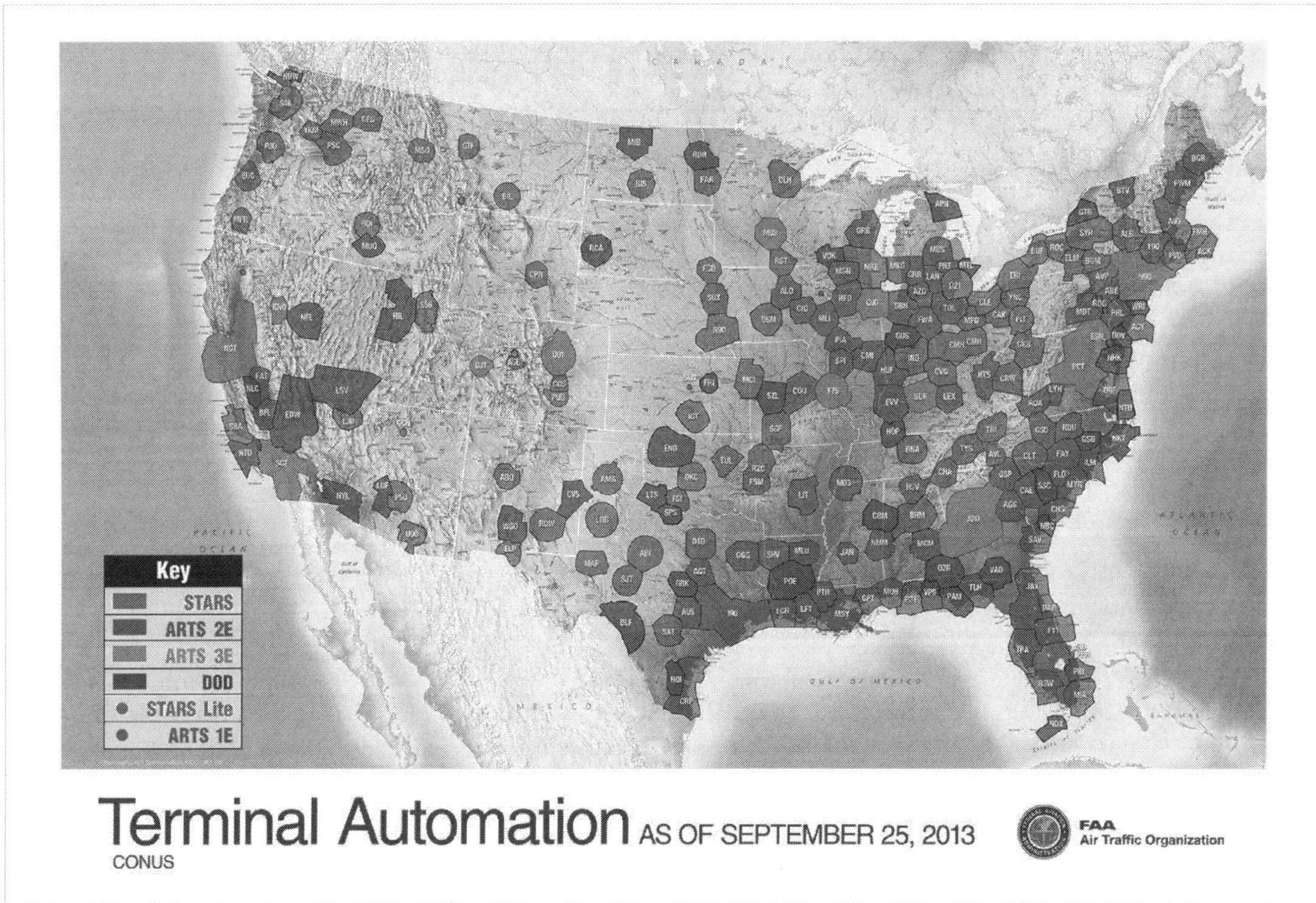

Figure 3-24. Computerized radar systems in use by the FAA.

Automated Radar Terminal System

One of the first systems installed by the FAA, designated ARTS-I, was installed in the Atlanta air traffic control tower in 1964. ARTS is designed to identify each aircraft by matching its transponder code with flight plan data. Once the aircraft is positively identified, the ARTS computer maintains constant identification and predicts the aircraft's future location. It also provides the controller with continuous alphanumeric information identification on the radar screen. This alphanumeric data includes an aircraft symbol and its associated data block. (See figure 3-25).

Figure 3-25. ARTS system display

The ARTS-I computer system proved to be highly successful in relieving controllers of the tedious task of maintaining the correct association between radar targets and flight progress strips. The successful implementation of ARTS in Atlanta and New York led the FAA to award a contract for the installation of additional, enhanced ARTS systems at high-activity airports across the United States. The current version of this advanced system is known as ARTS-IIIE and is now one of the standard automation systems for high traffic airports and consolidated approach controls.

ARTS-II

As traffic increased nationwide in the late 1960s and early 1970s, the FAA realized that a system of computerized radar processing would have to be developed for low to medium-activity facilities. The FAA determined that ARTS-III was inappropriate because it was too expensive to install and operate and required a team of computer programmers to maintain. In 1974 the FAA contracted with UNISYS to develop a low-cost ARTS radar processing system. To differentiate this system from the previous version of ARTS, it became known as ARTS-II.

The primary difference between ARTS-II and ARTS-III is that ARTS-II cannot track nontransponder-equipped aircraft. ARTS-II installations require little maintenance or routine programming, yet they are less flexible and less expandable. The FAA installed ARTS-II and its successor, ARTS-IIE, at over 100 smaller airports across the United States.

Standard Terminal Automation Replacement System

The standard terminal automation replacement system (STARS) is a newer terminal air traffic control system that uses modern commercial computing equipment to replace existing ARTS systems. It is a joint FAA/Department of Defense program to replace ARTS and other older technology systems at over 200 FAA and DOD terminal radar approach control facilities and associated towers. (See figure 3-26).

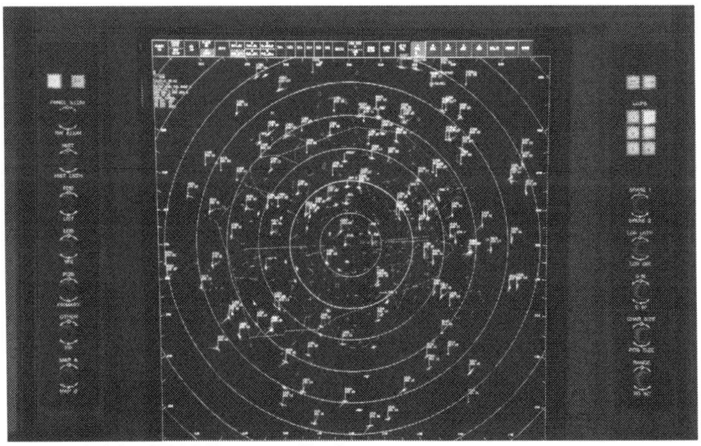

Figure 3-26. STARS system display

STARS includes 20x20-inch color displays with associated input devices, computers, and networks. STARS provides controllers with aircraft positional information with weather, flight data, and other ATC-related information overlaid directly on the display console. STARS uses commercial workstation computers connected to one another by local area networks to provide controllers with color displays that include controller-modifiable windows and graphics.

STARS uses a standard graphical user interface, thereby reducing the number of dials and knobs prevalent at conventional controller workstations. The system can easily be adapted by the programmer and/or the working controller. The workstations can also be easily reconfigured if sectors need to be combined or split. The STARS system is capable of tracking up to 1,350 aircraft simultaneously and can accept inputs from up to sixteen remote radar sites. This information can then be parsed out and supplied to more than 100 different controller positions in over twenty remote locations.

Enroute Automation Modernization

The enroute automation modernization (ERAM) system is the current system in place to display air traffic at the nation's ARTCCs. (See figure 3-27). ERAM is a modular system that provides controllers with all currently needed functionality but can easily be upgraded as the ATC system transitions into future systems. ERAM makes both radar data and flight plan information available to every controller. Controllers are able to share and coordinate information between centers and between centers and towers. ERAM can take multiple radar inputs and display the best data to each controller; this is known as a mosaic presentation. ERAM also has the capability to predict future aircraft trajectories. This capability can be utilized to provide the controller with advanced system capabilities such as conflict prediction and resolution, traffic management solutions and other, not yet implemented automation tools. Parts of the ERAM system also includes automatic information tracking of radar-identified aircraft, thus eliminating the need for the older manually updated flight progress strips controllers used to use for this purpose.

Figure 3-27. ERAM display

Chapter 4

Air Traffic Control Towers

There are many different duties performed by controllers in an air traffic control tower. Busier towers employ more controllers and divide up the duties differently than low-activity towers. The specific duties and responsibilities are tailored to each tower's or airport's needs and traffic conditions. Even very busy towers usually never staff every possible position all of the time; it depends on the amount of traffic and the airport configuration at any given time. All the possible positions in a control tower are listed in the table below. Unless the tower is unusually busy, some (or many) of the duties of each control position can be combined. At some low-activity towers, there might be as few as one or two controllers on duty at any given time, with many duties and responsibilities combined. The various control positions in towers are listed on Table 4-1.

TABLE 4-1 CONTROL TOWER POSITIONS

Designator	Control Tower Position
AD	Arrival Data (Radar)
AP	Approach control
AR	Arrival Control (Radar)
CC	Coordinator (Tower)
CC	Coordinator (Tower)
CD	Clearance Delivery
CI	Coordinator (Radar)
DC	Departure Control (Radar)
DD	Departure Data (Radar)
DR	Departure Control (Radar)
FD	Flight Data
GC	Ground Control
LC	Local Control
OM	Operations Manager
OS	Operations Supervisor
STMCIC	Supervisory Traffic Management Coordinator-in-Charge

Most control towers have at least three operating positions on duty at any given time, although sometimes the duties might be expanded to two or more controllers working similar positions if the workload is high enough. For example, a busy airport might have two local and two ground controllers working during peak times; this will generally only occur if the airport is operating parallel runways. Some airports have grown so large and busy that there are two or more separate control towers at the airport. In that case, the airfield and airspace are divided into different areas, with each area managed by a separate control tower.

In general, the local, ground, clearance delivery and flight data control positions are those most commonly staffed and will be discussed in this chapter. As each position is described, keep in mind that although the FAA has standardized duties and functions, each air traffic facility has its own unique requirements that might modify the generic job responsibilities described in this chapter.

Local Control

The local controller is primarily responsible for the separation of aircraft operating within the local traffic area, as well as those landing or departing on any of the active runways. The local controller is assigned a discreet radio frequency that permits communication with these aircraft. The primary responsibility of local control is arranging inbound aircraft into a smooth and orderly flow of traffic while sequencing departing aircraft into this flow. The local controller has authority over the active runway at all times; if aircraft must cross this runway during ground operations, the ground controller must obtain permission to do so from the local controller.

Specific duties performed by local control include:

- Separating aircraft on the same and intersecting runways including arrivals and departures.
- Issuing landing and takeoff clearances.
- Ensuring wake turbulence separation standards are maintained.
- Resolving emergency and other special situations.
- Coordinating with ground control for runway crossing/usage.
- Issuing weather and airport information to pilots including information obtained from Terminal Doppler Weather Radar (TDWR) or Low-Level Wind Shear Alert (LLWAS) systems if installed.
- Complying with any special terminal area local procedures, LOAs, facility directives, orders or notices.
- Operating the runway and approach light systems as well as any local radar display.

Local Controller Duties

One of the primary responsibilities of local control is to safely sequence arrivals and departures at the airport by ensuring that proper runway separation between aircraft exists. The local controller issues verbal instructions to arriving and departing aircraft to ensure this runway separation. It is not local control's responsibility to separate VFR aircraft while airborne and inbound to the airport, although the controller may offer assistance and issue traffic advisories. It is assumed that VFR pilots will apply the see and be seen rules of traffic avoidance when flying to and from the airport.

Runway Separation

The basic separation rule applied by local controllers is that an airplane will not be cleared to land or takeoff on any runway until any previous aircraft is clear of the runway. This rule also applies to intersecting runways. Once the runway is clear of the previous traffic, local control clears the next aircraft to land or to takeoff, whichever is appropriate. There are some exceptions to this basic rule however. For example, the controller does not have to necessarily wait until the first aircraft is clear of the runway prior to issuing a landing clearance to the second aircraft. According to the ATC handbook (7110.65), the controller can issue the clearance if he/she can reasonably assume that the separation will eventually exist. For instance, if an aircraft has just landed on a runway and is about to turn on to the taxiway, local control could issue a landing clearance to an aircraft that was two miles away from the runway, as there is a reasonable assurance that the runway will be clear when the second aircraft arrives. If for some reason it was not, the controller can simply rescind the landing clearance; this is what is known as a "go around".

There are some situations when a controller can permit two aircraft to use the same runway at the same time, but certain conditions must exist, and the procedures differ depending on whether both aircraft are landing, both departing, or one is departing and the other is landing. These procedures are called Same Runway Separation (SRS) rules; and are summarized below.

SRS Aircraft Categories

For the purpose of same runway separation (SRS), every aircraft is classified by the FAA into one of three SRS aircraft categories. The amount of separation required for each aircraft varies depending on its SRS category. The complete list of aircraft categorization can be found in the appendices to the Air Traffic handbook.

Category I aircraft are generally lightweight, single-engine, propeller-driven personal aircraft. This category includes aircraft such as the Beechcraft Bonanza series, the Cessna 152, 172, 182 and 210, Piper Cherokee and Arrow, Cirrus SR-20 and SR-22 and other similar small aircraft. It does not include high-performance single-engine aircraft such as military fighters or single engine jets. Helicopters are usually classified as category I aircraft.

Category II aircraft are lightweight, twin-engine, propeller-driven aircraft weighing 12,500 pounds or less. This category includes aircraft such as the Beech Baron and small (series 90) King Air aircraft, as well as most piston-engine Cessna and Piper twin engine propeller-driven aircraft. It does not include larger aircraft such as the Beech Super King Air series or the Douglas DC-3. Most turboprop driven aircraft are considered category III.

Category III includes all other aircraft not included in either category I or II. This category includes high-performance single-engine, large twin-engine, multi-engine propeller-driven, and virtually any jet powered aircraft. This includes prop driven aircraft such as the Douglas DC-3, Beechcraft King Airs and Beechjet, Learjets, Cessna Citations and other business jets as well as every Airbus, Bombardier (Canadair), Embraer and Boeing airliners.

Departing Aircraft Separation

The local controller is required to separate departing aircraft from one another by ensuring that an aircraft does not begin its takeoff roll until at least **one** of the following conditions exists:

1) The preceding departing aircraft is airborne and has crossed the departure end of the runway or has turned to avoid any conflict. (Figure 4-1).
2) Between sunrise and sunset, if local control can determine runway distance, the second aircraft can start its takeoff roll once the first aircraft is BOTH airborne AND the following minimum distance exists between the aircraft involved. (Figure 4-2).

   a) If both aircraft are SRS category I, 3,000-foot minimum separation is required.
   b) If a category I aircraft follows a category II, 3,000-foot minimum separation is required.
   c) If the second or both of the aircraft are category II, a 4,500-foot minimum separation interval is required.
   d) If any of the aircraft involved are a category III aircraft, a 6,000-foot minimum separation interval must be used.

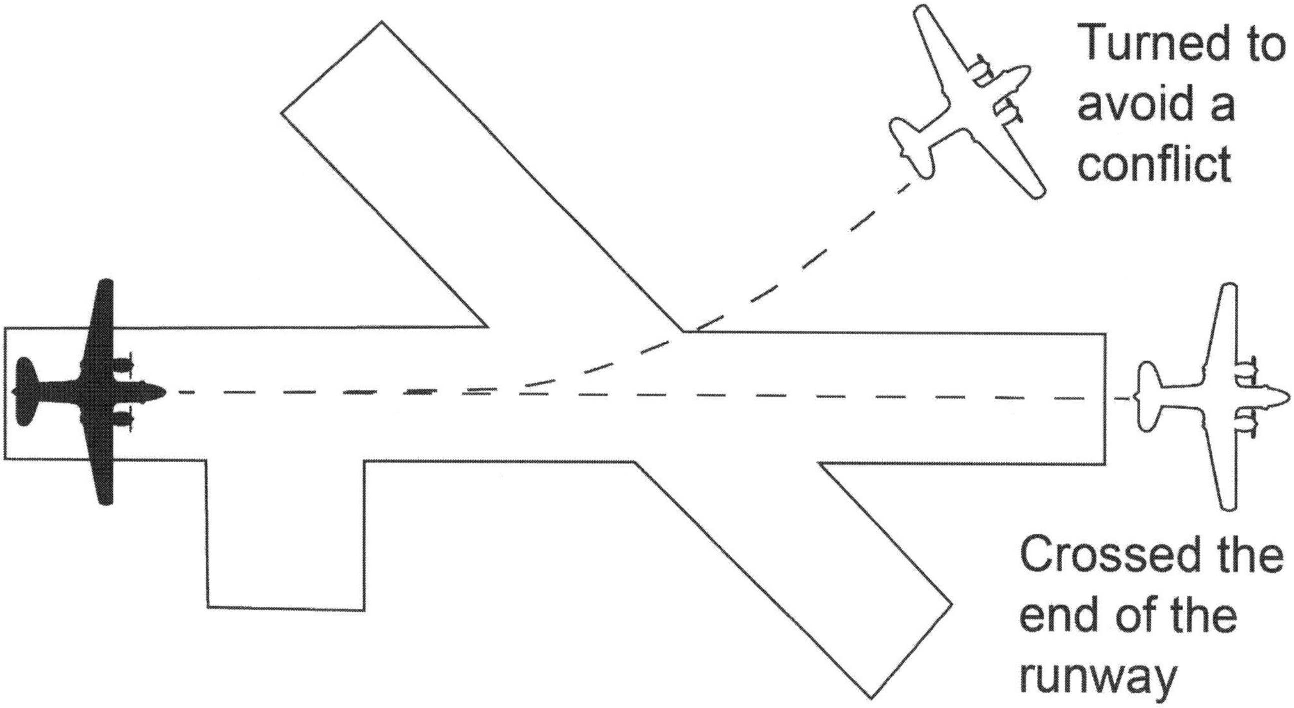

Figure 4-1. Departing aircraft separation

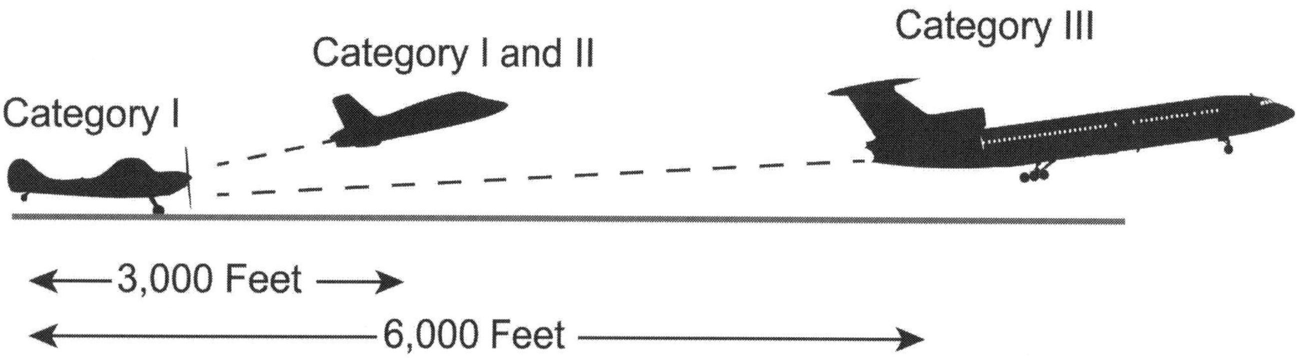

Figure 4-2. Departing category I aircraft using distance rule

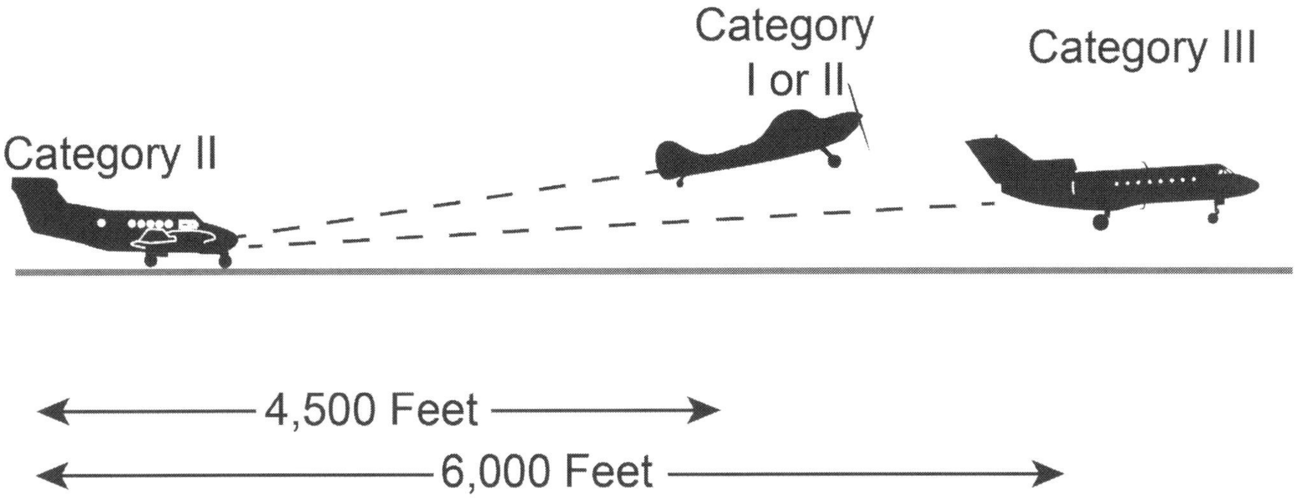

Figure 4-3. Departing category II aircraft using distance rule

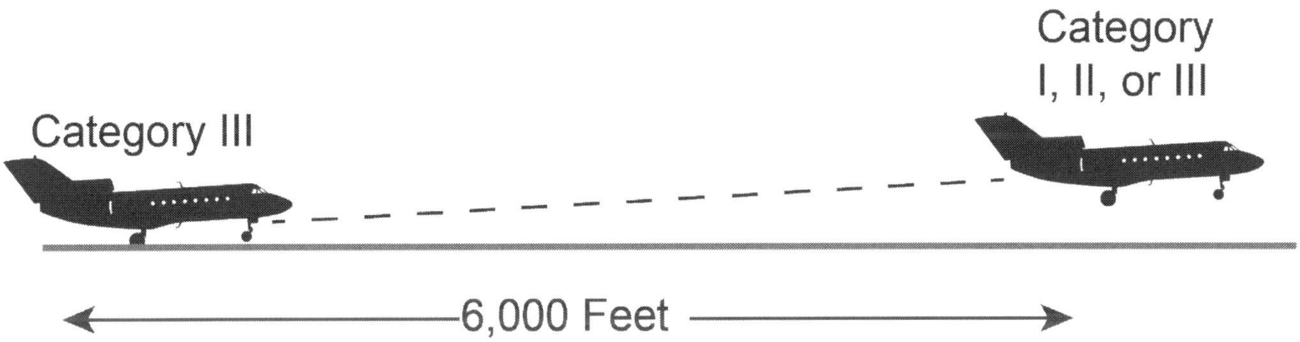

Figure 4-4. Departing category III using distance rule

Thus, if a Piper Cherokee (category I) departs and is followed by a Cessna 152 (also category I), local control must not permit the Cessna to begin its takeoff roll until the Piper has crossed the departure end of the runway, has turned to avoid a conflict, or is airborne and at least 3,000 feet down the runway.

But if the Piper is followed by a Cessna 310 (category II), local control must not permit the Cessna 310 to begin its takeoff roll until the Piper has crossed the departure end of the runway, has turned to avoid a conflict, or is airborne and at least 4,500 feet down the runway. If the Cessna 310 precedes the Piper, however, only 3,000 feet of separation would be needed.

But if the Piper is followed by a Boeing 737 (category III), the Boeing cannot begin its takeoff roll until the Piper has crossed the departure end of the runway, has turned to avoid a conflict, or is airborne and at least 6,000 feet down the runway. Even if the situation is reversed, (the Piper is following the Boeing), the Piper cannot begin its takeoff roll until the 737 has crossed the departure end of the runway, has turned to avoid a conflict, or is airborne and at least 6,000 feet down the runway.

Line Up and Wait

To increase runway utilization, it may be advantageous to have the second aircraft on the runway, in position to depart, waiting for the preceding aircraft to complete its departure. It may also be advantageous to have an aircraft holding, ready to depart on one runway while an intersecting runway is being used for either arrivals or departures. When this procedure is used, the pilot is advised to "line up and wait" (LUAW). The intent of LUAW is to position aircraft for an imminent departure. It is not to be used if the aircraft will be holding on the runway for any lengthy period of time. Whenever using this procedure, the controller must explicitly issue the clearance while advising the holding aircraft of the operation keeping the controller from authorizing an immediate takeoff. The phraseology to be used is AIRCRAFT ID, RUNWAY (number), LINE UP AND WAIT (traffic information). For example:

"United three twenty two, Lafayette tower, runway one zero, line up and wait, traffic departing runway two three."

"Cessna four six echo, Lafayette tower, runway three, line up and wait, traffic landing runway one zero."

Traffic information may be omitted when the traffic involved is another aircraft which has landed on or is taking off the same runway and is clearly visible to the holding aircraft. As part of the LUAW procedure, the "other" aircraft involved must be advised of the traffic holding on the runway as well.

"Delta one twenty, Lafayette tower, runway one zero, cleared for takeoff, traffic holding in position runway five."

"Air Shuttle sixteen fifteen, Lafayette tower, runway one zero, cleared to land, traffic holding in position runway two three."

Controllers should be careful when using a line-up-and-wait clearance to ensure that the pilot does not misinterpret the instruction as a takeoff clearance. It is for this reason that the word "cleared" is never used in conjunction with a clearance to line up and wait. Once the runway separation standards can be met, the instruction "cleared for takeoff" is issued to the second aircraft by local control. This instruction clears the pilot to perform a normal takeoff on the runway specified.

"United seven twenty-five, runway two three, cleared for takeoff".

Intersecting Runway Separation

If a departing aircraft is taking off on an intersecting runway that is in use for either landings or takeoffs, local control must ensure that an aircraft does not begin its takeoff roll until at least one of the following conditions exists:
The aircraft that has landed on the intersecting runway has either:

- taxied off the landing runway, or
- passed through the intersection, or
- completed the landing roll and advised local control that it will stop prior to the runway intersection.

If the aircraft operating on the intersecting runway is departing, that aircraft must have completed one of the following before a second aircraft can be cleared for takeoff from an intersecting runway.

- It must have crossed the intersection, or
- It must be airborne and turning prior to the intersection to avert a conflict.

Arriving Aircraft

Arriving IFR pilots usually fly a standard instrument approach that lines them up with the runway, whereas VFR pilots approach the airport using all or a portion of a standardized traffic pattern. (A typical left-hand traffic pattern is shown in Figure 4-5). It is local control's responsibility to properly space these two types of inbound aircraft while also sequencing departures into the traffic flow.

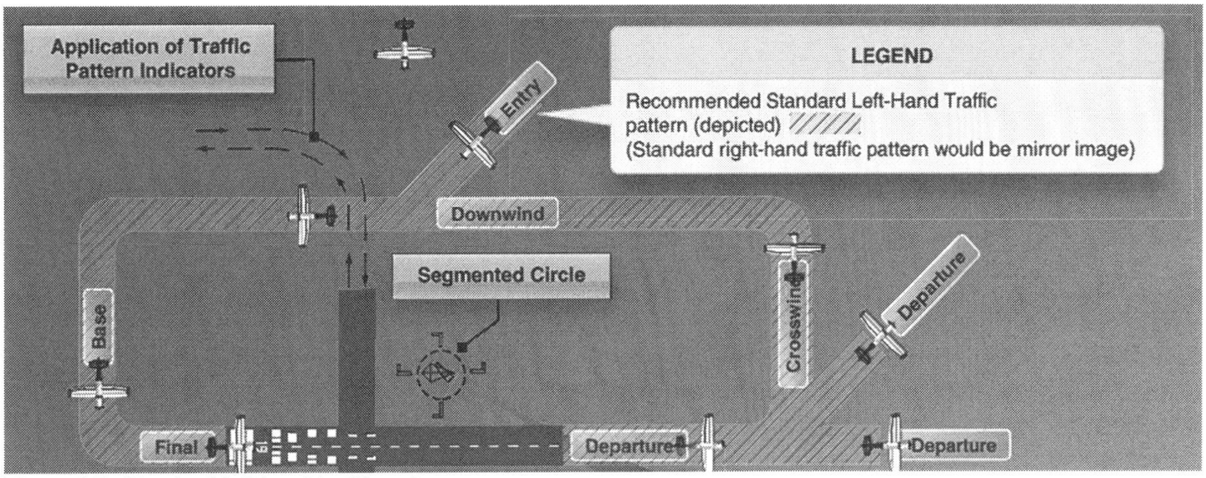

Figure 4-5. Standard left hand traffic pattern

A VFR traffic pattern consists of five portions known as traffic pattern legs.

- **UPWIND** is the flight path parallel to the landing runway in the direction of landing and departure. It is normally used primarily for departing aircraft but might be used for aircraft conducting multiple practice landings and takeoffs (touch and goes).
- **CROSSWIND** is the flight path at a right angle to the landing runway on the departure end but heading away from the airport.
- **DOWNWIND** is the flight path parallel to the landing runway in the direction opposite of landing.
- **BASE** is the fight path leg at a right angle to the landing runway off its approach end extending from the downwind leg to the intersection of the extended runway center line.
- **FINAL** is the flight path leg in the direction of landing along the runway centerline and extends from the base leg to the runway threshold.

If the turns performed by the aircraft in the pattern are to the left (as shown in figure 4-5), the traffic pattern is known as left traffic. The crosswind, downwind and base legs are respectively referred to as left crosswind, left downwind and left base. If all turns in the pattern are made to the right, it is known as right traffic. Unless specified in facility directives, either left or right traffic can be used for any runway at a tower-controlled airport; left traffic is considered standard at uncontrolled airports.

Sequencing and Spacing of Aircraft

It is local control's job to organize the traffic in the pattern into a safe and logical flow (sequencing). It is also their job to space the aircraft appropriately to ensure that there will be adequate runway separation for both arrivals and departures. There are many ways to accomplish this. The controller can instruct the aircraft to enter the traffic pattern at different points or even place aircraft in different traffic patterns (left vs. right).

If needed, the controller can ask the pilot to modify the traffic pattern, stretching it out or shortening it. Aircraft can also be asked to fly circles (360 degree turns) while in the pattern to provide additional spacing.

The local controller typically uses some of the following phraseology to achieve proper sequencing of aircraft in the traffic pattern:

- **"Enter (pattern leg) runway (runway number)".** The controller uses this phrase to direct the pilot to enter one of the five identified pattern legs. For example, "Cessna niner papa uniform, enter left downwind runway two three."

- **"Report (position)".** The pilot can be requested to make various position reports. The controller may request distance from the airport, distance from the runway, distance from a prominent landmark, or entry into the pattern. "Diamond eight delta mike, report three miles north of the airport". "Cherokee two papa uniform, report over the red and white water tower". "Cirrus zero two romeo, report two mile final runway one zero".

- **"Number (sequence number, runway)".** This advises the pilot of the planned landing sequence for the aircraft. This instruction is usually used in conjunction with a "follow" phrase.

- **"Follow (description and location)."** The local controller should advise the pilot of the location and type of any preceding aircraft to make it easier for the pilot to locate and follow. "Diamond one seven three, number two, traffic is a Cessna on a quarter mile final runway one zero".

Spacing Instructions

The following instructions can be issued by a local controller to either increase or decrease the spacing between aircraft in the traffic pattern.

- **"Extend Downwind/Upwind".** The pilot can be requested to extend either the downwind or the upwind leg a specified distance or until over a prominent landmark. "United six sixteen, extend downwind one mile". "Cessna one niner foxtrot, extend upwind to the lake".

- **"Short Approach".** A short approach is a request for the pilot to shorten the downwind leg as much as possible, which results in an equivalent reduction in the length of the final approach leg. "Cherokee five pap uniform, make short approach runway one zero".

- **"Make Left (or Right)".** In a normal traffic pattern, the pilot makes a 90° turn when transitioning from one traffic pattern leg to another. One method of increasing the spacing between two aircraft is to request that the pilot turn 270° in the "wrong" direction when transitioning to the next leg. For instance, if the pilot is on a right downwind, a request to "make a left two seventy to base" will result in a longer turn and increased separation. If the pilot is not transitioning from one leg to another and increased spacing is necessary, a 360° turn in either direction may be requested. "Cessna one two romeo, make a right three sixty".

Landing Clearance

Once the aircraft is in a position to land, and it can be reasonably assumed that the required separation will be obtained and maintained, a landing clearance can be issued. There are a number of landing clearance variations that can be used by local control.

"Cleared to land" authorizes the pilot to make a full-stop landing. "Cessna two six mike, runway two three, cleared to land". After the aircraft has landed, local control advises the pilot where to exit the runway and what frequency to use to contact ground control.

A touch and go clearance permits an aircraft to land on the runway, but then take off again before actually coming to a stop. This maneuver is usually used by students practicing takeoffs and landings. An aircraft performing a touch and go is considered an arriving aircraft until actually touching down and then is considered a departure. "Cirrus four papa uniform, runway one zero, cleared touch and go".

Go Around

If it is apparent that proper runway separation cannot be achieved and neither aircraft's traffic pattern can be adjusted, it may be necessary to cancel landing clearance for one of the aircraft. In this case, local control determines which aircraft's landing clearance should be canceled and instructs that aircraft to "go around." Upon receipt of this instruction, the pilot immediately begins a climb back to pattern altitude and reenters the traffic pattern as instructed. "American six eleven, go around, enter right downwind runway two seven left".

Arrival Runway Separation

The local controller is required to apply runway separation standards to arriving aircraft just like departures. This requirement is accomplished by requiring the pilots to adjust their flight pattern as necessary to provide the required separation. If only one runway is in use, local control separates aircraft from one another by ensuring that a landing arriving aircraft does not cross the runway threshold until at least one of the following conditions exists:

- The preceding aircraft has landed and taxied off of the runway.
- Between sunrise and sunset, the preceding aircraft need not have taxied off of the runway if the distance between the two aircraft can be determined using landmarks or runway markings, and the following minimums can be maintained:
  - A distance of 3,000 feet if a category I aircraft is landing behind either a category I or a category II aircraft (Figure 4-6), or
  - A distance of 4,500 feet if a category II aircraft is landing behind either a category I or a category II aircraft (Figure 4-7).

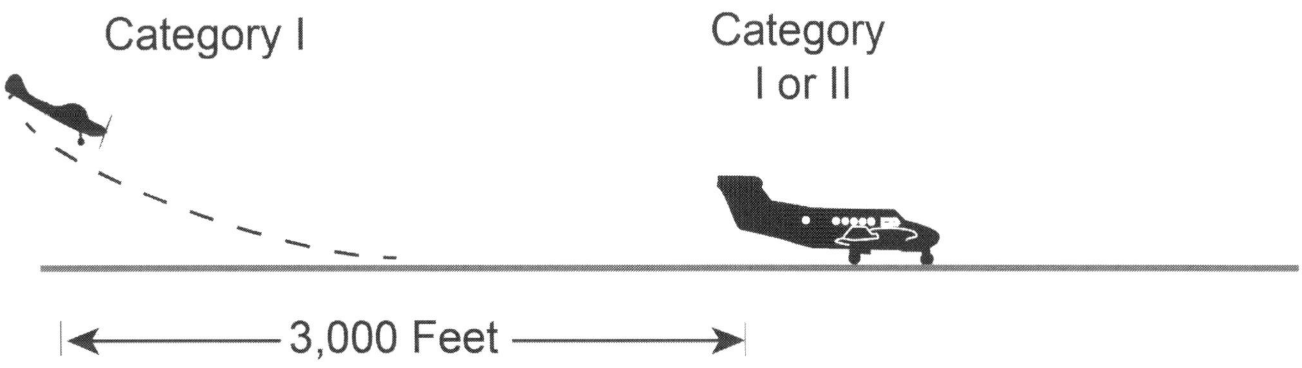

Figure 4-6. A category I aircraft landing on the same runway behind a category I or II

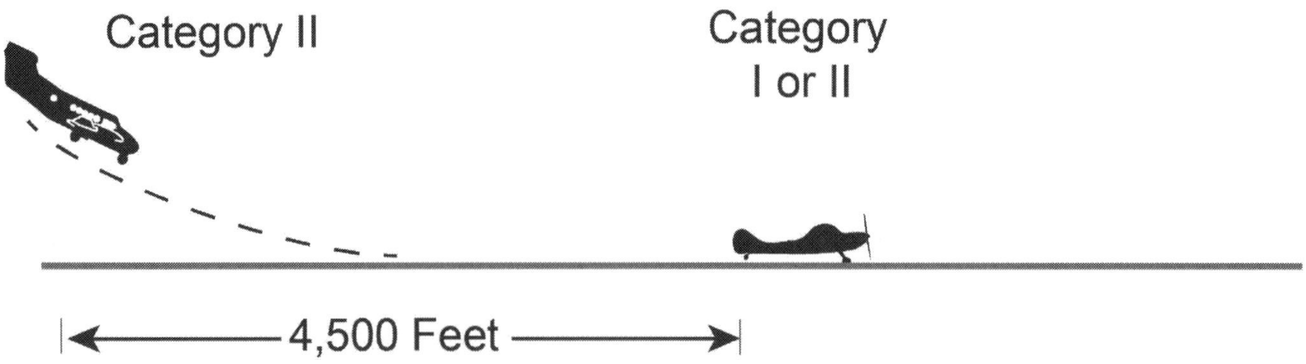

Figure 4-7. A category II aircraft landing on the same runway behind a category I or II

Arrival-Departure Separation

If an aircraft wishes to depart the same runway behind an arriving aircraft that has just landed, takeoff clearance cannot be issued until the landing aircraft has taxied off of the runway. If a landing aircraft is following a departure however, the controller must insure that the departing aircraft has crossed the departure end of the runway before the arrival crosses the runway threshold. This requirement can be disregarded if, during daylight hours, the first (departing) aircraft is airborne and is at least the following distance down the runway from the landing threshold:

- 3,000 feet if a category I aircraft is landing behind a departing category I or II aircraft (Figure 4-8), or
- 4,500 feet if a category II aircraft is landing behind a departing category I or II aircraft. (Figure 4-9), or
- 6,000 feet if either of the aircraft involved (landing or departing) is a category III aircraft (Figure 4-10).

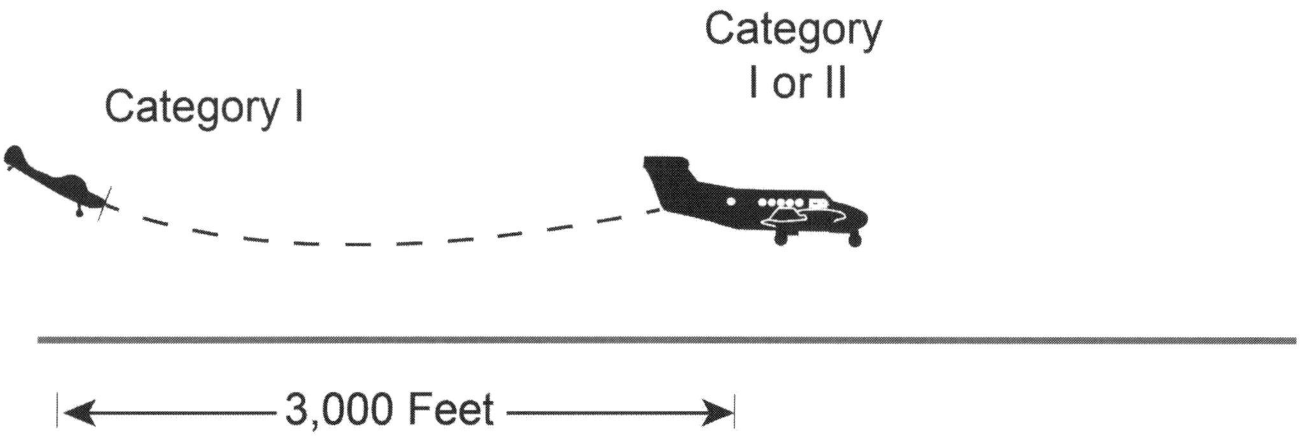

Figure 4-8. A category I aircraft landing on the same runway behind a departing category I or II

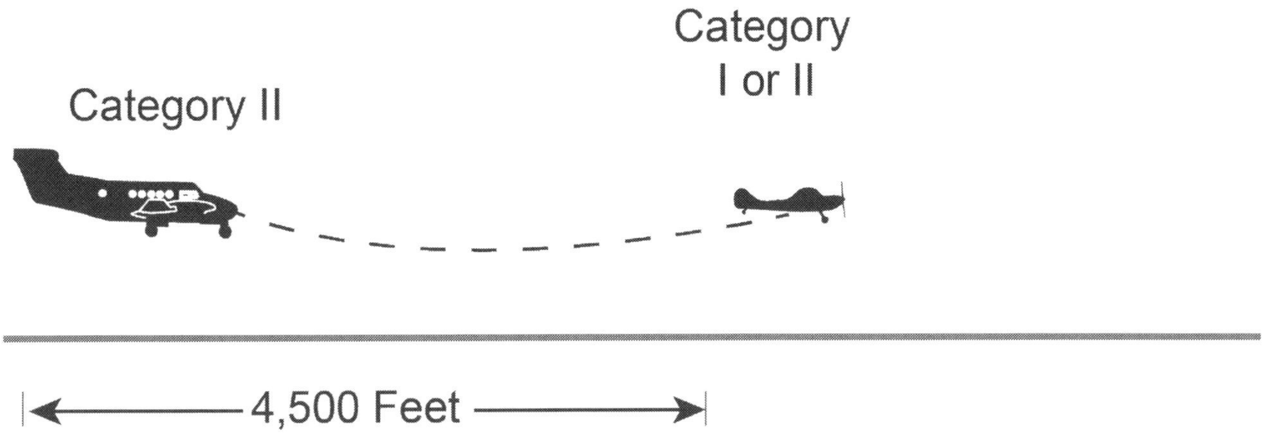

Figure 4-9. A category II aircraft landing on the same runway behind a departing category I or II

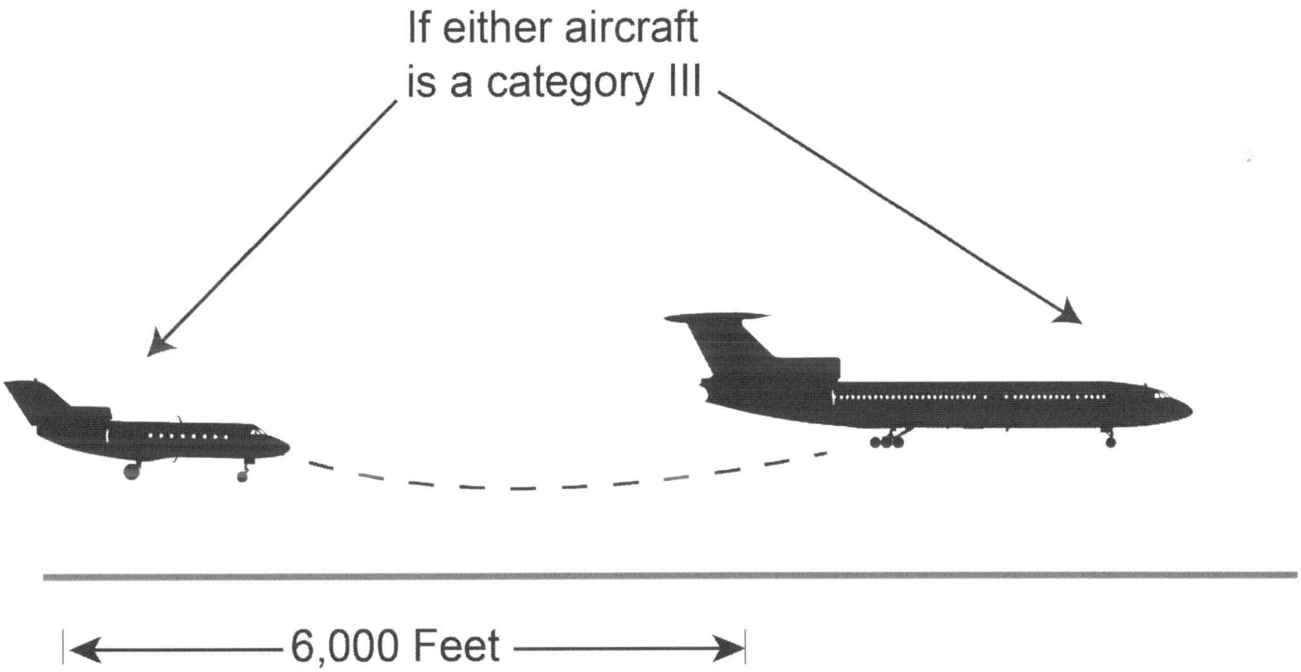

Figure 4-10. An aircraft landing on the same runway behind a departing aircraft when either (or both) are category III

Intersecting Runway Separation

If intersecting runways are in use, an aircraft landing on one runway must be sequenced so as not to cross the landing threshold until at least one of the following conditions exists:

- If an aircraft is departing from an intersecting runway it must have either crossed the intersection or turned to avert any conflict.
- If an aircraft is landing on the intersecting runway, it must have either turned off the landing runway, crossed the runway intersection, or completed the landing roll and advised local control that the aircraft will hold short of the intersecting runway. (Figure 4-11).

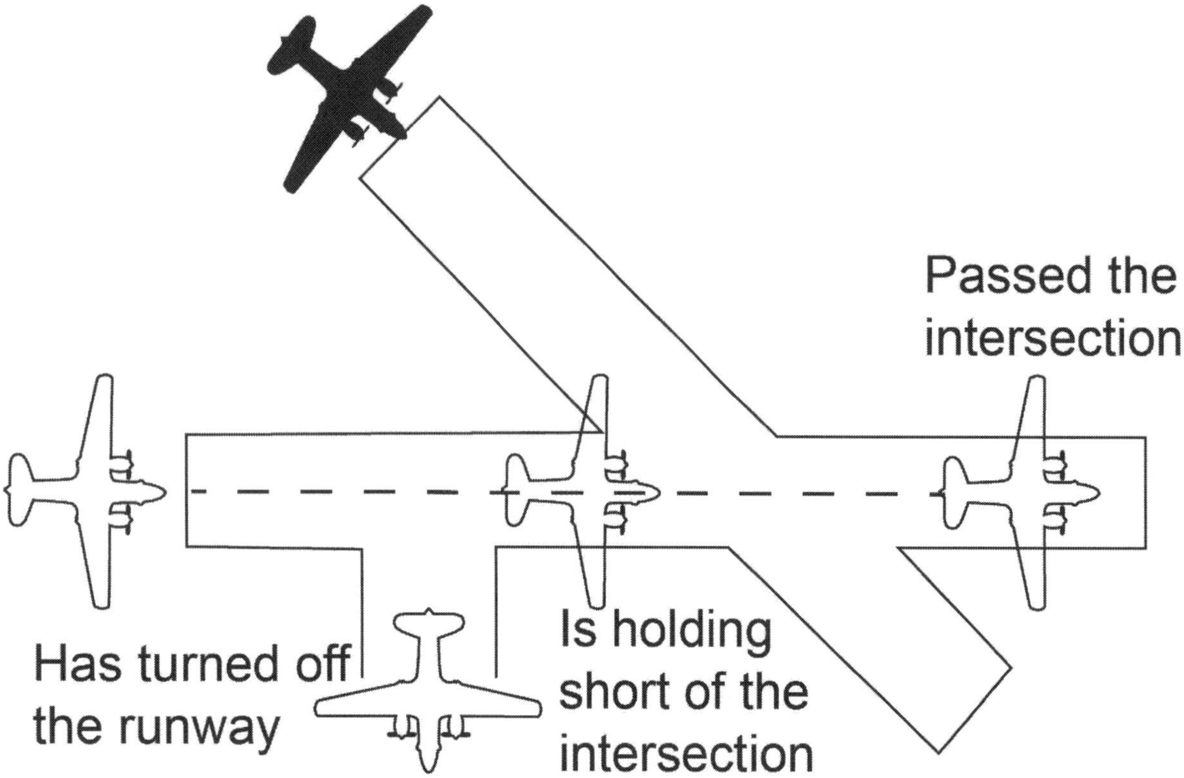

Figure 4-11. Intersecting runway separation

Land and Hold Short Operations

At certain select airports, air traffic controllers may use an intersecting runway separation procedure whereby one aircraft can land and hold short of an intersecting runway, while another aircraft operates on that runway. This operation is known as a land and hold short operation (LAHSO). Special procedures have to be developed and approved and appropriate runway or taxiway signage has to be installed before controllers can issue LAHSO clearances. LAHSO procedures improve the efficiency of certain airports by essentially eliminating crossing runways. Although the runways in question may in reality cross each other, by requesting that a pilot hold short of the intersection, the controller can act as if the runways were physically disconnected.

Pilots may accept LAHSO clearances provided they determine that their aircraft can safely land and stop within the available landing distance (ALD). ALD data are published and made available to the pilots or controllers can provide the same data to the pilot upon request. (Figure 4-12).

LAFAYETTE

PURDUE UNIVERSITY (LAF)(KLAF) 2 SW UTC–5(–4DT) N40°24.74′ W86°56.21′ **CHICAGO**

606 B S4 **FUEL** 100LL, JET A OX 1, 2 TPA—1606(1000) Class IV, ARFF Index A **H–5E, 10F, L–27D**

NOTAM FILE LAF **IAP, AD**

RWY 10–28: H6600X150 (ASPH-GRVD) S–90, D–110, 2S–140, 2D–165 HIRL

RWY 10: MALSR. PAPI(P4L)—GA 3.0° TCH 50′. Trees.

RWY 28: REIL. VASI(V4R)—GA 3.0° TCH 46′. Trees.

RWY 05–23: H4225X100 (ASPH) S–48, D–58, 2D–90 MIRL 0.3% up NE

RWY 05: REIL. PAPI(P2L)—GA 3.0° TCH 38′.

RWY 23: REIL. PAPI(P2L)—GA 3.0° TCH 39′. Thld dsplcd 323′. Fence.

LAND AND HOLD–SHORT OPERATIONS

LDG RWY	HOLD–SHORT POINT	AVBL LDG DIST
RWY 10	05–23	4600

RUNWAY DECLARED DISTANCE INFORMATION

RWY 05: TORA–4225 TODA–4225 ASDA–4225 LDA–4225

RWY 23: TORA–4225 TODA–4225 ASDA–4225 LDA–3912

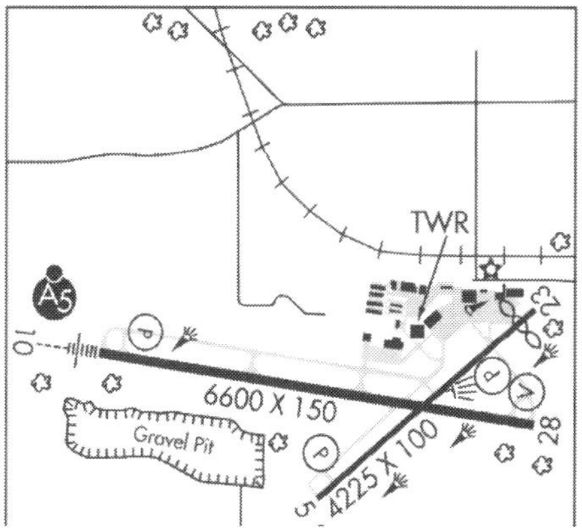

Figure 4-12. LAHSO runway length information

Although it is up to the controller to initiate a LAHSO clearance, the pilot has the final authority to accept or decline any land and hold short clearance. The safety and operation of the aircraft remain the responsibility of the pilot. LAHSO operations are only authorized when all the following conditions can be met:

- A ceiling of 1,000 feet and a visibility of at least 3 miles exists at the airport.
- The runway is dry.
- The tailwind on the LAHSO runway is calm or less than 3 knots.
- LAHSO markings, lights and signage are installed (Figure 4-13).

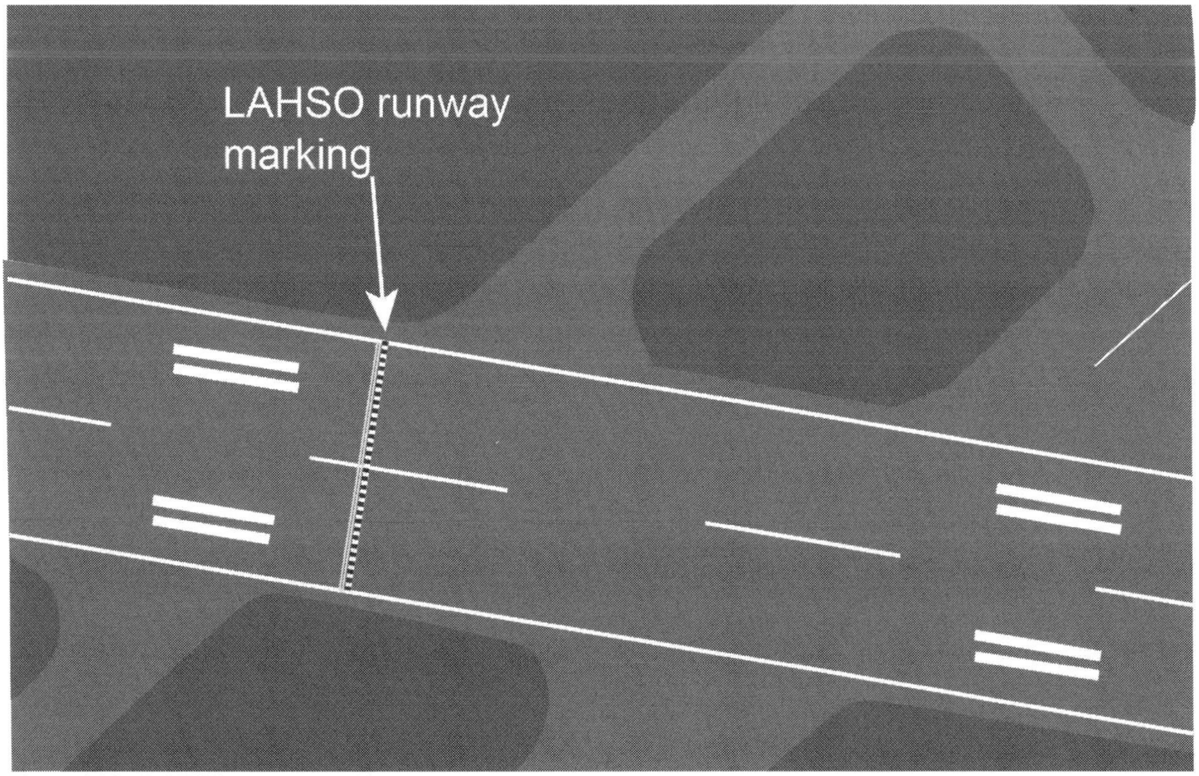

Figure 4-13. LAHSO runway markings

Wake Turbulence

Behind every aircraft in flight there is an area of unstable air, known as wake turbulence. This turbulence was originally attributed to "prop wash" but is now known to be caused in part by a pair of counter-rotating vortices trailing from the wing tips (Figure 4-14). These vortices are a by-product of the lift produced by the wing, which is generated by the creation of a pressure differential between the lower and the upper wing surfaces. High pressure is created below the wing, while low pressure is created above. The resultant upward pressure on the wing, known as lift, causes whirling vortices of airflow to be created at the wing tip. The airflow along the wing pushes this upward flow backward, creating a whirling body of air that resembles a horizontal tornado.

Figure 4-14. Wing tip vortices

Each wing produces its own vortex, resulting in two counter-rotating cylindrical vortices trailing from each aircraft. The strength of the vortex is governed by the weight, speed, and shape of the wing. In general, maximum vortex generation occurs when the aircraft is heavy and slow--precisely the conditions found during takeoff and landing. Wing-tip vortices created by larger aircraft can completely encompass smaller aircraft. The rotational velocities in these vortices have been measured in excess of 100 knots, causing a small aircraft encountering one of these vortices to become completely uncontrollable.

Wing-tip vortices begin to be generated the moment an aircraft's nose wheel lifts from the ground and are continually created until the aircraft lands (Figure 4-15). These vortices tend to descend at 500 feet per minute until they level off at about 900 feet below the aircraft's cruising altitude. They remain at this point until dissipating. If they contact the earth's surface while descending, they tend to move outward at a speed of about five knots. Any surface wind will tend to help dissipate and move these vortices. A crosswind will tend to increase the speed of the downwind vortex, while impeding the progress of the upwind vortex (Figure 4-16). A cross wind between 3 and 7 knots may prevent the upwind vortex from moving at all, sometimes leaving it hovering over the runway for a minute or two!

Figure 4-15. Wake turbulence generation

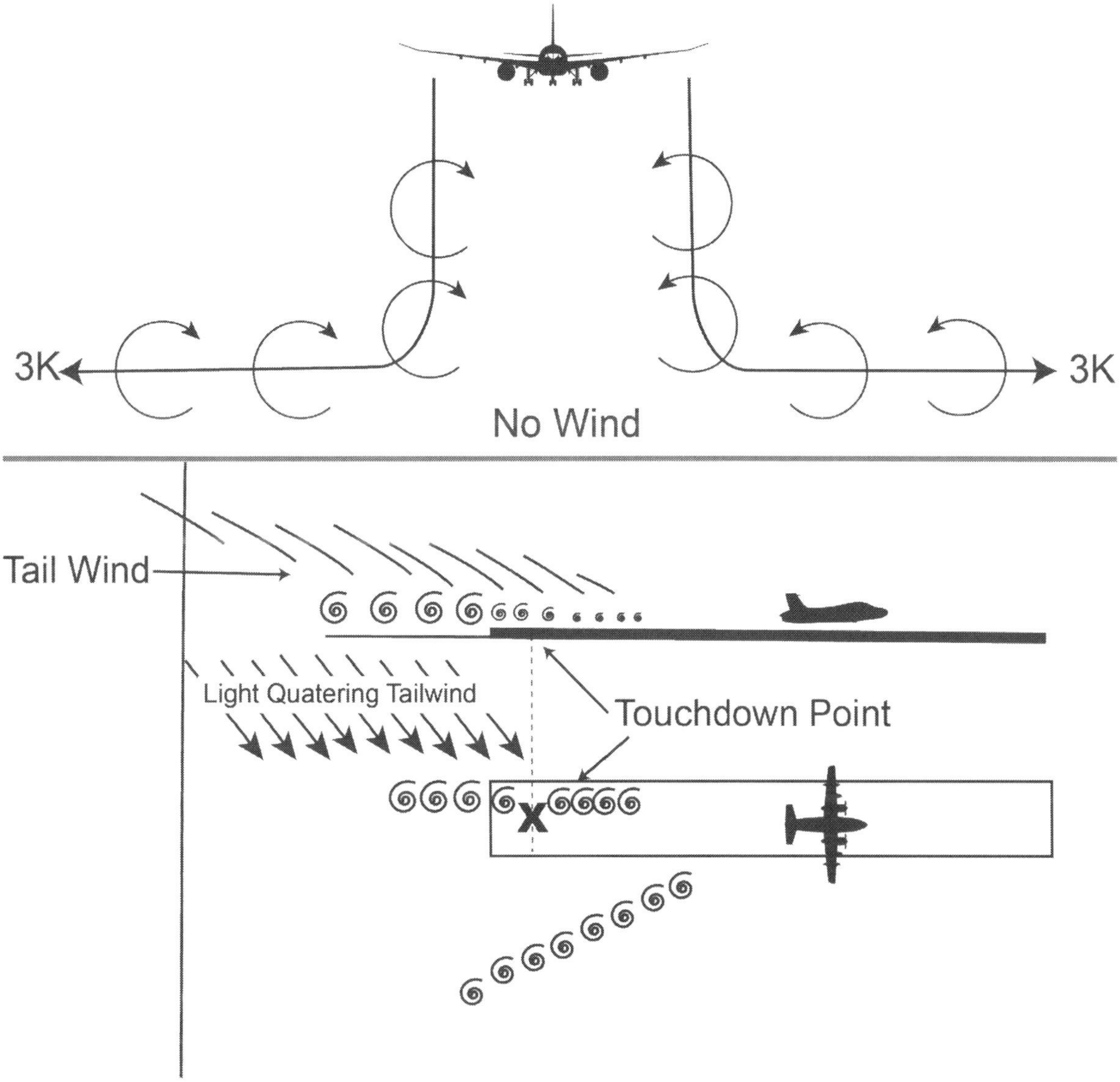

Figure 4-16. Vortex movement across the ground

Wake Turbulence Weight Classes

Although it's the pilot's responsibility to avoid wake turbulence, controllers are required to assist pilots of smaller aircraft whenever they fly behind an aircraft potentially creating dangerous wake turbulence. To help these pilots, the FAA has assigned every aircraft to one of three weight classes; small, large, or heavy.

Small aircraft are aircraft whose maximum certificated takeoff weight is less than or equal to 41,000 pounds. Large aircraft have maximum certificated takeoff weights greater than 41,000 pounds up to and including 300,000 pounds. Heavy aircraft have maximum certificated takeoff weights in excess of 300,000 pounds. A weight class of Super is used for Airbus A380 and Antonov AN225.

Since wake turbulence tends to dissipate in a matter of minutes, time is used to ensure that a following aircraft does not encounter severe wake turbulence. In general, aircraft are usually delayed by either a two- or three-minute interval wherever dangerous wake turbulence might exist. Two minutes of separation is applied to any class aircraft when departing behind a heavy aircraft using the same runway (or a parallel runway if the runways are separated by less than 2,500 feet.) Two minutes of separation must also be applied to an aircraft whose flight path will cross that of a heavy jet departing from an intersecting runway. Three-minute separation is provided to any small aircraft departing behind a large aircraft whenever the small aircraft is departing from an intersection or in the opposite direction on the same runway.

To help pilots better identify aircraft that could create significantly larger amounts of wake turbulence, the word "heavy" is appended to the call sign of all aircraft in the heavy weight class. Recent introduction of very large aircraft has required a modification of the wake turbulence rules. Both the Airbus A380-800 (A388) and Antonov An225 (An225) create wake turbulence in excess of that created by a "heavy" aircraft. These two aircraft are therefore referred to as "super" as part of their call signs and require that the controllers increase spacing behind these two aircraft. The usual separation for operations behind a Super aircraft is to delay the takeoff clearance to the following aircraft until at least three minutes have passed since the preceding aircraft (the Super) began its takeoff roll. There are a lot of slightly different wake turbulence separation rules that take into account intersecting and opposite direction runway landings and takeoffs that need to be applied. Not every airport will have each of these configurations however.

Ground Control

The ground controller is responsible for the separation of aircraft and vehicles operating on the ramp, taxiways, and any inactive runways. This responsibility includes aircraft taxiing out for takeoff, aircraft taxiing in after landing, and any ground vehicles operating on airport movement areas. Airport movement areas do not include those areas solely reserved for vehicular traffic such as service roads or boarding areas.

The ground controller is assigned a unique radio frequency to communicate with pilots and vehicle operators. The most common ground control frequency is 121.90 MHz. In congested areas where two or more control towers are located near each other, ground control transmissions from each airport might overlap, causing pilot misinterpretation. In these rare cases each control tower is assigned a different frequency for its ground controllers. These additional frequencies are usually 121.80 or 121.70 MHz

The ground controller issues taxiing instructions to the pilot using the radio. These instructions include the aircraft identification, the name of the ground controller's facility, the route to be used while taxiing, and any restrictions applicable to the pilot. Some examples of potential ground clearances follow:

> "United six eleven, Lafayette ground, runway one zero, taxi via taxiway Alpha."
> "American nine twenty-one, taxi to the terminal via taxiways charlie and delta one."

Preventing Runway Incursions

One of the primary responsibilities of ground control is to ensure that vehicles and taxiing aircraft remain clear of the active runways. If an aircraft or vehicle must cross an active runway, ground control must receive permission for that operation from local control. If an aircraft should inadvertently taxi onto an active runway without local control's knowledge, an accident could result. Such accidental entry, known as a runway incursion, should be avoided at all costs.

One of the best ways to prevent a runway incursion is to use and understand the appropriate phraseology for communicating with taxiing aircraft. Runways have been assigned identifying numbers (runway three-two, or two-seven left for example). Taxiways are all identified by letters (such as A, B, C, etc.). This verbal identification of runways and taxiways reduces the chance of a misunderstanding leading to a runway incursion.

If an aircraft is required to taxi across an active runway ground control must coordinate with local control to receive permission to cross runway. If permission is not received, ground control must advise the pilot to stop prior to the runway. This is known as holding short of the runway. The ground controller does not have to coordinate the crossing of any inactive runway as an inactive runway is usually assigned to ground control for their use.

To help insure that pilots don't mistakenly taxi across an active runway, pilots are taught that they cannot cross any marked runway without express permission to do so. This requires that ground control expressly clear the aircraft to cross every runway. This is the best way to ensure that pilots don't inadvertently cross a runway without ground control's knowledge.

> "JetBlue twenty-three eleven, Anchorage ground, runway two five left, taxi via taxiway charlie, hold short of runway two five right."
> "United four twelve eleven, Anchorage ground, cross runway three three."

Normally only one runway crossing clearance may be issued at a time. If two or more runways need to be crossed, progressive hold instructions followed by individual runway crossing clearances are usually issued. At certain facilities, if two runways are so close together that it would cause a hazard to have an aircraft stop between the two while awaiting a crossing clearance, the FAA will establish procedures that permit ground control to issue multiple runway crossing clearances in a single radio transmission.

Flight Data

The flight data controller assists the other controllers in the tower and performs any needed clerical duties. This is the position typically assigned to a new controller at the facility. (Figure 4-17). The basic responsibilities of and duties performed by a flight data controller include the following:

- Receiving and relaying IFR departure clearances to the clearance delivery controller
- Operating the flight data processing equipment
- Relaying weather and NOTAM information to other operational positions
- Aiding other tower controllers by relaying any directed information
- Collecting, tabulating, and storing daily records
- Preparing the Automatic Terminal Information Service (ATIS) recordings
- Processing field condition reports

The flight data controller is also required to disseminate appropriate weather information to other controllers or to the National Weather Service (NWS). Tower controllers may might solicit or receive pilot reports

(PIREPs) from pilots operating within the vicinity of the control tower. PIREPs are an essential means of passing along actual flight conditions to other pilots and the NWS. The flight data controller is responsible for disseminating this weather information to pilots through the use of ATIS equipment.

ATIS is a continuous-loop digital recording, usually made by the flight data controller and transmitted on a VHF frequency for pilot reception. ATIS recordings inform both arriving and departing pilots of weather conditions and other pertinent information at the airport. Pilot reception of ATIS information relieves the ground or approach controller of repeating weather conditions and non-control information to every aircraft, which would increase frequency congestion. Recordings are made at least once every hour but may be made more often if weather conditions change rapidly.

The following information is included in an ATIS:

- The name of the airport.
- The ATIS phonetic alphabet code.
- The UTC time of weather observation.
- Wind direction and speed.
- The visibility in miles and/or fractions of a mile.
- The cloud ceiling measured in feet above the ground.
- The temperature in degrees Celsius.
- The dew point temperature in degrees Celsius.
- The altimeter setting.
- The instrument approach procedure(s) currently in use.
- The runways(s) used for arrivals.
- The runway(s) used for departures.
- Pertinent NOTAMS or weather advisories. These include any taxiway closures, severe weather advisories, navigation aid disruptions, unlit obstacles near the airport, bird or wildlife information, or any other problems that could affect the safety of flight.
- Braking action reports (if appropriate).
- Low-level wind-shear advisories (if appropriate).
- Remarks or other information. This may include VFR arrival frequencies, radio frequencies that have been temporarily changed, runway friction measurement values, bird activity advisories, and part-time tower operation.
- Instructions for the pilot to advise the controller that the ATIS recording has been received.

A typical ATIS recording would sound like this:

"Lansing Airport information charlie, one five five zero zulu weather, wind one six zero at one zero, visibility five, light snow, measured ceiling six hundred broken, two thousand overcast. Temperature seven, dew point two, altimeter two niner five five. ILS runway two eight left approach in use, landing and departing runways two eight left and two eight right. Notice to airmen, taxiway delta is closed. VFR arrivals contact Lansing approach control on one one eight point six five. Advise the controller on initial contact that you have information charlie."

Figure 4-17. Clearance Delivery/Flight Data controller

Clearance Delivery

At most control towers, the clearance delivery controller performs the flight data tasks in addition to those assigned to clearance delivery. The tasks specifically assigned to clearance delivery include issuing clearances to departing aircraft and amending clearances as needed.

Clearances for departing IFR aircraft are generated by the ARTCC based upon the flight plan filed by the pilot or their company. At Class B and C airports, VFR aircraft receive an abbreviated departure clearance that ends once they leave the class B or C airspace and enter class E (or G) airspace.

Clearances normally include the following:

- Aircraft identification
- Clearance limit
- Departure procedure
- Route of flight
- Altitude
- Departure frequency
- Transponder code

Aircraft clearances should always be given in the same order. Once the clearance delivery controller verifies the pilot has accurately recorded the clearance and makes any required modifications, the clearance is issued to the pilot using a discreet radio frequency or through a computer interface. If any changes need to be made, the clearance delivery controller accomplishes this using a computer tied into the aircraft clearance database. This clearance instructs a pilot what to do after departure; it does not allow an aircraft to depart at an airport with a tower. The aircraft will still need to receive clearances for taxi and departure from other terminal controllers.

Datalink Services

The FAA is phasing in the use of digital communications with the aim of replacing verbal communications conducted between the aircraft and clearance delivery. Many high traffic airports are now equipped with the Tower Data Link System (TDLS) that includes the Pre-departure Clearance (PDC) and Digital ATIS (D-ATIS) functions.

The PDC function automates the clearance delivery operations for qualifying participating users. PDC digitally transmits the aircraft's IFR clearance via a data link directly to the aircraft. PDC is available only to participating aircraft that have subscribed to the service through an approved service provider. No acknowledgment of receipt or read back from the pilot is required for a PDC. It is assumed that the pilots have received the clearance unless the pilot contacts clearance delivery and states otherwise. In conjunction with the PDC program, the FAA has established airports where the ATIS can also be received digitally by the aircraft in a manner similar to PDC.

Flight Progress Strips

After issuing the IFR clearance to the pilot, clearance delivery passes the relevant aircraft information to ground control. This information is printed in a standard format on forms known as flight progress strips (Figure 4-18). To facilitate accurate interpretation, flight strips are printed using standard markings and abbreviations, ensuring that specific information will always be found in the same place. These locations on the flight progress strip are known as fields. The approved field contents and format can be found in the Air Traffic Control handbook.

Figure 4-18-1. Terminal flight progress strip.

3 1 2 4 5 6 7 8 9 10	11 12 13 14 14a	15 16 17 18 19	20 20a	21 22 23 24	25 26	27 28 29 30
DAL542 1 H/B753/A T468 G555 16 16 486 09	7HQ 1827	30 18 PXT	330 RA↑1828		FLLJ14 ENO 000212 COD PHL	2675 *ZCN

Figure 4-18-2. ARTCC Flight progress strip

Flight progress strips used in control towers are formatted differently from those used in the ARTCCs but contain essentially the same information. Traditionally, controllers have written flight information and amended clearance directly on the flight progress strip, which then gets passed along to the next controller. Paper flight progress strips have been replaced in many facilities with digital displays that contain the same information (Figure 4-19). Controllers amend digital flight progress strips via a computer keyboard and/or mouse input. Electronic flight strips can be modified for display depending on their use but generally mimic the format commonly found on paper strips.

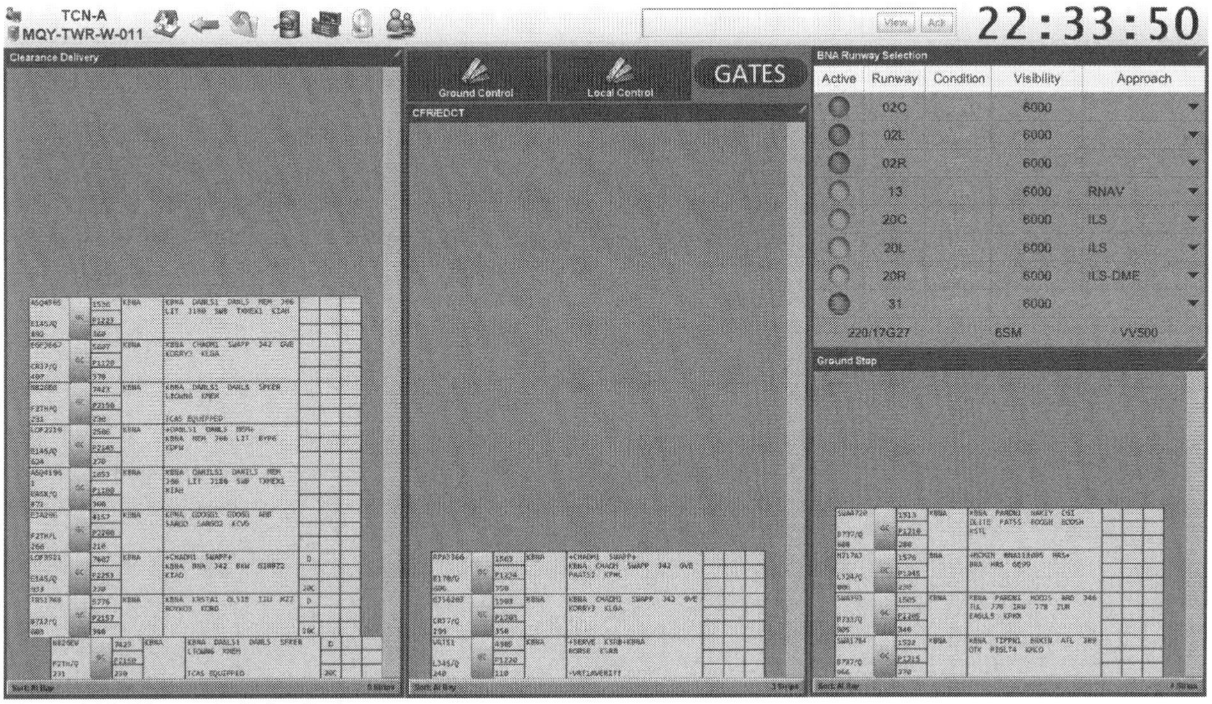

Figure 4-19. Electronic flight display system

The format for flight strips differs somewhat depending on whether the aircraft involved is a departure, arrival, or an overflight (an aircraft that passes through the airspace delegated to the tower but is not landing). Flight progress strips typically contain the following information:

- Aircraft identification.
- Flight strip revision number
- Type of aircraft. A heavy aircraft is identified with an H preceding the aircraft type on the flight strip. An equipment suffix is added to the aircraft type to let the controllers know the aircraft's navigational equipment.
- Assigned transponder code.
- Time. For a departing aircraft, this is the proposed departure time (UTC) that the pilot filed in the original flight plan. For an arrival or overflight aircraft, an estimated time over a nearby fix will be provided, as well as the arrival time if appropriate.
- Altitude. This is either altitude requested in the pilot's flight plan (for a departure) or the aircraft's current altitude (for an arrival or overflight). To conserve space on the flight progress strip, the last two zeros in the altitude are dropped.
- Departure and/or arrival airport. This is the airport from which the aircraft will depart or to which it will arrive. Every airport that has a published instrument approach has been issued a four letter identifier for use in flight planning. Some systems and users drop the first letter "K", as all airports in the continental U.S. use "K" as the first letter of their identifier. Some of the more common identifiers include:
    - KORD--O'Hare International, Chicago, Illinois
    - KJFK--John F. Kennedy International, New York
    - KATL--Hartsfield International, Atlanta, Georgia
- Route of flight. The route to be flown includes any airways or VORs that the pilot will be using. If the route is to be flown using area navigation (RNAV), either the waypoint names or their latitude-longitude coordinates will be included. If no airway is designated between two VORs, it is assumed that the pilot will fly directly from one VOR to the next.

- Any preferential routes that have been assigned by the ARTCC computer. A preferential route may be a departure, an enroute, or an arrival route.

Chapter 5

Approach Control

The FAA operates over 150 radar approach control facilities across the U.S. (See figure 5-1). Some are associated with control towers; others are larger, "consolidated" facilities responsible for one or more busy airports. Each of these facilities normally controls airspace up to about 10,000'-15,000' above the ground out to a distance of approximately 40 miles from the primary airport. Within this airspace, controllers are required to separate IFR aircraft from one another. If the TRACON's airspace includes class B, C or D airspace, VFR aircraft may be separated as well. In general, the separation rules and operating procedures described in this chapter are those primarily applied to IFR aircraft.

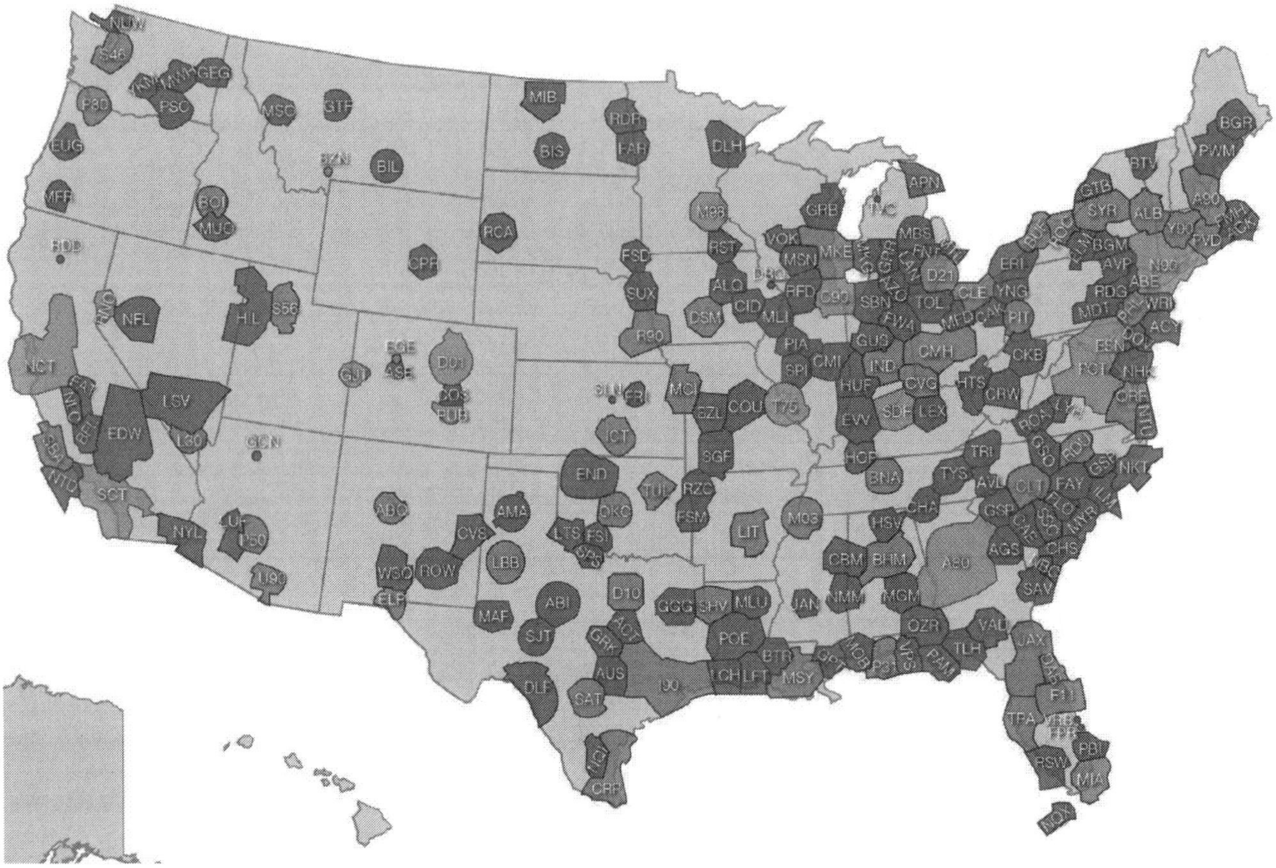

Figure 5-1. TRACON airspace in the U.S.

Basic IFR Separation

Before the widespread installation and use of radar for air traffic control, controllers were unable to accurately determine the location of the aircraft they were attempting to separate. In most cases, separation was accomplished by instructing the pilot to change course or altitude, or to enter a holding pattern to remain well clear of other traffic. Direct radio communication between the controller and the pilot was not always possible, with communication often passed through intermediaries, such as airline radio operators. The resultant delay, (which could be as long as 30 minutes), further complicated the controller's task.

Figure 5-2. RCAG site.

Some of this delay was alleviated through the development and use of remote radio transmitters and receivers known as remote communication air/ground (RCAG) devices. RCAG permits controllers to communicate with pilots whose aircraft are beyond the range of the radio transmitters at the control facility. (Figure 5-2). RCAG units use telephone circuitry connected to remote radio transmitters and receivers. But even with the help of RCAG, controllers still had to rely on pilot reports and handwritten flight progress strips to separate aircraft.

Flight Progress Strips

Fight progress strips became an invaluable tool to help controllers to separate aircraft. Through using standardized procedures and markings, controllers visualized the relative position of every aircraft to apply proper separation. Whenever the controller issued an instruction to a pilot or the pilot made a position or altitude report, the controller wrote this information on the flight progress strip. The strip could then be easily interpreted by others to determine the status of the aircraft. The flight progress strip also became a valuable record of instructions issued by the controller and reports made by the pilot in the case of an investigation following an incident or accident. It was the controller's responsibility to continuously obtain and update information contained on the flight progress strip to be able to visualize both the present and, more importantly, the future position of every aircraft being separated.

Controllers use these strips to maintain knowledge of where aircraft are and where they are going to be; they then use standards established in the controller handbook to determine if a safe distance exists between aircraft. Unlike the tower operations discussed in Chapter 4, TRACON controllers cannot see the aircraft themselves, so they rely on some form of the radar systems previously discussed. They then use what they know about the aircraft's location and route to ensure that the "safe, orderly and expeditious flow of traffic" is maintained.

As technology changes, the FAA relies on air traffic controllers to ensure these standards are maintained as they refine procedures when new equipment is available. They must consider many variables to ensure that aircraft that the new equipment and procedures will actually improve ATC capabilities and aviation safety. Previously, these variables might include ground-based and airborne navigation equipment errors, different navigation systems in use, winds aloft, and problems caused by delayed communications. Currently, as things like satellite communications and cyber security become more prevalent, controllers must remain vigilant to ensure that new equipment and procedures improve the efficiency and safety of the overall system. They must also consider worst-case scenarios thus ensuring that even with the maximum possible error in each component of the system a sufficient margin of safety is still maintained.

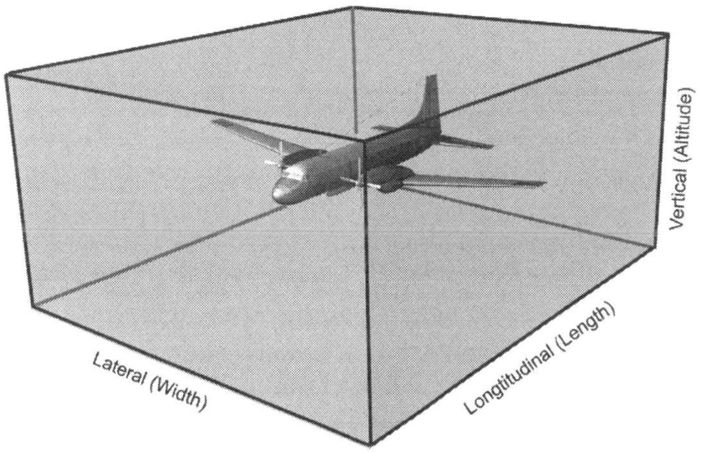

Figure 5-3. Reserved airspace for IFR aircraft.

With the above in mind, it is necessary to note that when every variable is taken into account, aircraft may not be as precisely located as their pilots or controllers think they are, even with the enormous improvements in aircraft monitoring. For that reason, controllers are trained to think of an aircraft occupying a block of airspace, not a single point; this is known as their protected airspace. The controller's job then is to ensure that no other aircraft infringes upon this airspace; if they did, it would be an operational error. The size of this airspace block is partially determined by the variables mentioned earlier, as well as other factors such as aircraft performance and altitude, navigation system used and distance from navaids/radar. Separation is also dependent on the speed and direction of the aircraft. While separating aircraft, controllers must assume that an aircraft could be located anywhere within their block of airspace and must then separate every block of airspace accordingly. (Figure 5-3). Radar controllers then (those at TRACON or enroute) use some form of radar to determine where the aircraft is presently, which in turn determines where the protected airspace lies. Tower controllers, contrastingly, use visual information to ascertain an aircraft's position; once determined however, they rely on this same concept of a block of protected airspace to separate aircraft. Areas of reserved airspace may come into contact with another, but they must never be allowed to overlap, since any overlap might permit two aircraft to get too close to another. The separation of each aircraft's reserved airspace is the only way to ensure that the aircraft remain safely separated.

Radar Separation Procedures

Although both TRACON and enroute controllers can use either radar or nonradar for separation, TRACON controllers primarily focus on radar separation and techniques. A discussion of nonradar separation is included in Chapter 6. Radar separation is routinely much more efficient than nonradar separation and can also be utilized to provide additional ATC services to pilots. In general, radar is used by controllers to provide the following service for pilots:

- Aircraft separation
- Navigation assistance
- Instrument approaches
- Traffic advisories

Radar Identification

Before using radar to provide separation, the controller must positively identify the aircraft in question. Since radar can display hundreds of targets at any given time, the controller must be absolutely certain of a particular aircraft's identity prior to offering radar service to that pilot. If, for instance, a controller provided navigational assistance to the wrong aircraft, the resultant incorrect instructions might cause the aircraft to crash into terrain or into another aircraft.

An aircraft can be positively identified using either the primary or the secondary radar system. Primary radar identification is usually employed when the aircraft is not equipped with a transponder or when the transponder is inoperative or operating intermittently. The easiest method for identifying an aircraft using primary radar is to observe an aircraft that has just departed from an airport. Since only one aircraft can depart from a runway at any given time, it can be safely assumed that a departing target observed within 1 nautical mile of the departure end of the runway is positively identified. In order to use this rule, the radar controller needs to have a verbal or nonverbal notification of departures, which they can get from the tower controller.

A second method of identification using primary radar requires the pilot to report over an exact location known by the controller and displayed on the radar screen; this is known as position correlation. Possible locations that could be used include airway intersections, navigational aids, prominent terrain features, or nearby cities or towns. If the controller observes a single target located over one of these landmarks, positive identification of that aircraft can be presumed.

The third method of identifying a primary target using radar is to request that the pilot turn the aircraft a set amount of degrees (30 degrees or more is required) and then observe which blip on the radar performs the proper turn. Using this method generally requires at least two turns, as the controller will be required to get the aircraft back on their original heading. When using this method, controllers also need to take terrain avoidance into consideration.

More accurate methods of positively identifying aircraft can be provided by the secondary surveillance radar system. Because of its accuracy, in most situations it is preferable to use these methods to determine an aircraft's identity. The most common means of secondary surveillance radar identification employed by TRACON facilities uses the IDENT feature included in the transponder. When the pilot presses the IDENT button, the transponder transmits a special identification pulse (SIP) to the receiver on the ground, which interprets the SIP and causes the radar to display a double-width beacon slash on the radar display. Since the special identification pulse is the only method by which this double-width slash is typically produced, the use of the IDENT feature provides accurate, positive identification of an aircraft.

If the controller is using a radar system that can display the aircraft's transponder code or call sign directly on the display, the controller can positively identify an aircraft by requesting that the pilot select a specific transponder code and verifying that the display indicates correctly. This is accomplished by requesting that the pilot "squawk" a particular code (e.g., "Delta six twenty-two, squawk two one four five").

A third method of radar identification utilizes the standby function of the transponder. In this method, controllers instruct an aircraft to "squawk standby"; when the aircraft complies, their beacon target disappears. The controller will then pause briefly to ensure the correct aircraft disappeared, and then instruct the aircraft to "squawk normal". Doing so will cause the aircraft target to reappear.

The final method of secondary radar identification is the automatic acquisition method. When an aircraft squawks the proper beacon code, the aircraft information data block will then automatically associate (or acquire) that target and display information directly on the radar display. If this correlates with the reported position of the aircraft, positive identification of the aircraft is accomplished. Presently, this method is only available for controllers at enroute centers.

Transfer of Radar Identification

Once an aircraft has been radar identified by a controller, radar can be used to provide separation. It is usually to everyone's advantage to use these procedures as they are more efficient and provide the pilot and controller with more flexibility. As aircraft fly from one control sector to another, successive controllers need not repeat any of the radar identification procedures so long as the radar identification of the aircraft has been continuously maintained and transferred to each subsequent controller. If an aircraft's radar identification is not transferred, controllers must use nonradar techniques to separate that aircraft until its radar identification can be reestablished. This procedure of transferring identification and control is known as a "handoff".

Handoffs and Pointouts

Because the transfer of radar identification is so important to the operation of the ATC system, the FAA has developed standardized procedures and terminology to be used when transferring aircraft identification. A "handoff" is the process by which one controller (the transferring controller) transfers the radar identification and control of an aircraft to another controller (the receiving controller), radio communications with that aircraft will subsequently also be transferred. A "pointout" is used when a controller wishes to transfer the radar identification of an aircraft to another controller but retain control of that aircraft; radio communications with that aircraft will also be retained by the transferring controller. In either case, radar identification must be transferred from one controller to the next before the aircraft can traverse the boundary between air traffic control sectors. An aircraft is not permitted to cross a boundary between two sectors without the knowledge and the permission of the receiving controller.

If a handoff is being conducted, the transfer of communication must be accomplished before the aircraft crosses the sector boundary. This permits the receiving controller to be in radio contact with the pilot prior to the aircraft crossing the sector boundary. The receiving controller may not issue any clearance that will change the aircraft's route of flight or altitude while it is still within the transferring controller's sector. The receiving controller must approve the handoff before the aircraft crosses the sector boundary. Normally, this should be accomplished through automated handoffs using the computerized radar system. However, if that process is unsuccessful, the transfer of communication can be accomplished through verbal communication (typically accomplished using intrafacility intercom or leased telephone circuits). If verbal communication is required, the following format should be used:

- For a handoff: handoff, location, aircraft identification, altitude
    - "Handoff, over ORD, Alaska 43, flight level 280
- For a pointout: pointout, location, aircraft identification, altitude, routing
    - "Pointout, over MSP, Delta 92, one -five thousand, direct Des Moines"

Radar Separation

Radar is used by controllers as a supplemental tool to separate aircraft—it does not completely replace nonradar separation procedures. It does, however, permit a reduction of lateral and longitudinal separation minima, thereby increasing the efficiency and effectiveness of the controller. There are still many occasions when a radar-equipped facility might use nonradar separation in lieu of a radar procedure. In some cases, nonradar methods may be easier to apply and do not significantly restrict the pilot or reduce ATC system efficiency. In other areas, radar coverage may not extend as far as the FAA would like, and nonradar procedures are still used to separate aircraft. In addition, some FAA air traffic control facilities remain unequipped with radar and must rely solely on nonradar procedures.

But in most cases, use of radar increases ATC system efficiency, reduces controller workload, and enhances safety. When using radar, controllers can track the position of each aircraft, permitting most separation standards to be reduced. Radar also permits the controller to issue headings to pilots to more effectively use the airspace. The routine use of holding patterns has been virtually eliminated in the TRACON environment by using radar. Lost aircraft can be assisted, pilots can be warned of nearby traffic, and instrument approaches can be conducted solely through the use of radar. Radar can also be used to more efficiently manage airspace by lining up aircraft for approaches, and timing (metering) their entry or exit into specific areas.

Radar separation criteria are defined in much the same way as nonradar separation criteria. Specific procedures and criteria are used when applying vertical, lateral, longitudinal, or initial separation of aircraft. As when using nonradar procedures, the controller needs to apply only one of these methods of separation to any particular aircraft.

Vertical Separation

Vertical separation criteria for radar controllers is similar to that used by nonradar controllers. Most of the time, 1000' of reported altitude difference between aircraft provides vertical separation. This minimum separation does change with altitude however, based on the inherent accuracy of aneroid alitimeters.

Aircraft must be vertically separated (either nonradar oar while using radar) by the following minima:

TABLE 5-1 VERTICAL SEPARATION

Altitude	Required separation
000-FL410	1000 ft. (Special equipment required above FL290)
FL410-FL600	2000 ft
Above FL6000	5000 ft

An exception can be used when two aircraft are either climbing or descending. In such instances, the following aircraft can be assigned the altitude vacated by the previous aircraft once the pilot reports leaving that altitude, or if the controller observes a valid indication on radar or via voice communication that the first aircraft has passed through that altitude.

Longitudinal and Lateral Separation

Nonradar longitudinal and lateral separation depends on the many other factors including the aircraft's speed and course and is rarely used in TRACON airspace. When using radar for separation, the basic longitudinal and lateral separation criteria is 3 nautical miles (nm). Since the width of radar pulses increases as they travel away from the antenna, distant targets appear much larger on a radar display than those located closer to the radar antenna. For this reason, the FAA requires increased separation (5 nm) for aircraft located more than 40 nm from the radar antenna. (Figure 5-4). A table of radar separation follows:

TABLE 5-2 LONGITUDINAL AND LATERAL SEPARATION

Type of usage	Radar separation standard
Within 40 miles of the radar site	3 nm
More than 40 miles of the radar site, below FL600	5 nm
Any altitude, above FL600	10 nm

Visual Separation

One of the most flexible means of separating aircraft is through the use of visual separation. In general, visual separation requires that one of the pilots sees the other aircraft or that the controller is able to observe both aircraft and assume the responsibility for providing separation. Visual separation may be employed so long as radio contact is maintained with at least one of the aircraft and at least one of the following conditions can be met:

- The pilot of one aircraft has been informed of the traffic, can visually identify the other, and has accepted responsibility for separation. The pilot of the second aircraft has been informed that visual separation is being applied if their paths are likely to converge.
- The controller can visually identify both aircraft and can keep them separated by issuing verbal instructions. This form of visual separation can only be used by tower controllers.

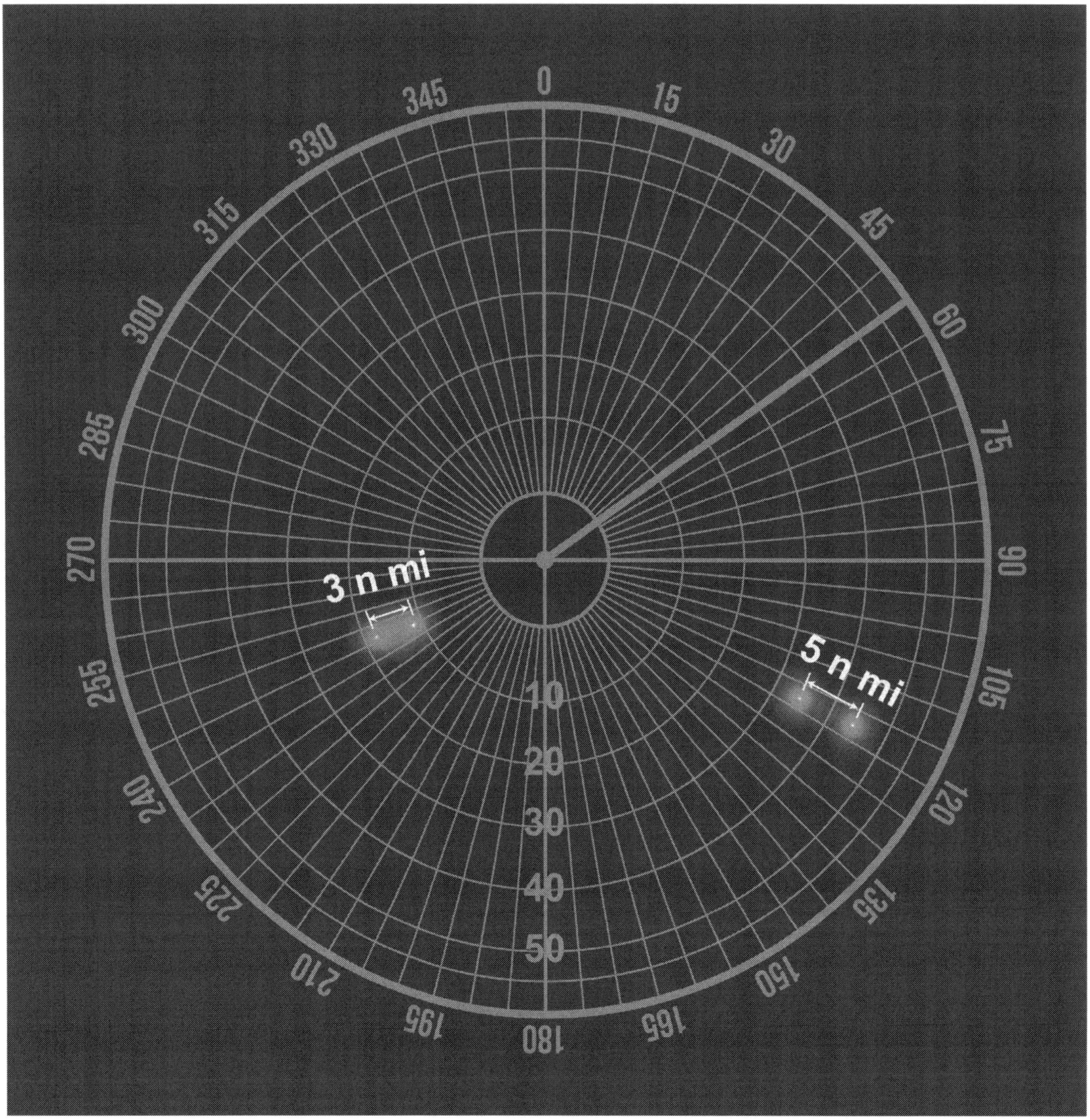

Figure 5-4. Lateral and longitudinal separation using radar.

Radar-Assisted Navigation

One of the advantages of using radar is that controllers can provide navigational assistance to pilots, potentially eliminating "dog legs" in the flight path. Rerouting an aircraft using radar also permits the pilot to bypass congested areas, thereby reducing or eliminating the use of holding patterns while enroute. Additionally, controllers can use radar to guide aircraft around potentially dangerous weather. Finally, radar can be used to position aircraft directly on the final approach segment of an instrument approach procedure, eliminating the need for airspace and time-consuming procedure turns; by doing so, controllers can significantly decrease the number of delays at busy airports.

Controllers assisting pilots to navigate issue verbal heading instructions to the pilots known as vectors. When vectoring an aircraft, the controller instructs the pilot to either turn to a specific magnetic heading, turn left or right a specific number of degrees, or simply to fly a particular heading. Here are some examples of the phraseology used for issuing vectors:

"Cirrus five papa uniform, turn left heading three-five-zero"
"Brickyard thirteen thirty three, turn right heading zero-niner-zero"
"Air Shuttle six forty five, fly heading one-three-zero"
"United seventeen, fly present heading"
"Beech zero two romeo, turn twenty degrees left"
"Delta sixteen ninety two, turn thirty degrees right"

Radar Arrivals and Approaches

Radar can be used to expedite arrivals to the final segment of an instrument approach. Instead of requiring each aircraft to conduct a lengthy procedure turn or enter a holding pattern before transitioning to the approach (two of the techniques used in a nonradar environment), a radar controller can vector each aircraft directly onto the final approach course. Radar vectors to final permit a controller to more effectively manage the spacing of aircraft with dissimilar flight characteristics.

In general, when vectoring aircraft for an approach, the radar controller issues instructions to place the aircraft in an extended traffic pattern. Since every aircraft (and operator) have different operating characteristics, the controller must continuously monitor the spacing of the aircraft. Generally, the goal is such that when one aircraft lands and taxis off the runway, there is another aircraft on close final. This spacing maximizes the use of the runway but requires the radar controller to take into account the changing speed and altitude of each aircraft. The controller also has to take into account changes in wind speed and direction, other airport restrictions (such as crossing runways), or the need to provide a "gap" in the arrival flow for departures on the same runway.

Traffic Flow in TRACON Airspace

At most busy airports, enroute controllers begin to arrange aircraft into a proper sequence about 200 miles from the airport. The enroute controllers issue speed restrictions to keep the aircraft spaced while descending the aircraft. Each Tracon has a Letter of Agreement (LOA) with the ARTCC that describes the spacing, sequencing and procedures that are to be used. Generally, the LOAs require that the ARTCC controllers begin to sequence the arriving aircraft such that they enter approach control's airspace descending to an altitude of about 10,000'-15,000' msl while flying over one of several designated arrival fixes, known as arrival gates.

Most large airports procedurally separate inbound and outbound aircraft using a modification of a "box and cornerpost" system of separation. In a typical configuration, the LOA describes a box drawn around the affected TRACON's airspace. Each corner of the box, known as a cornerpost, is delineated by an intersection or navaid. For example, at Chicago the cornerposts are the Krena, Kubbs, Plano and Bearz intersections. At Indianapolis, the cornerposts are delineated by the Jells and Antti intersections to the northwest, Clang to the northeast, the Shelbyville (SHB) VOR to the southeast, and the Kelly intersection to the southwest. (Figure 5-5).

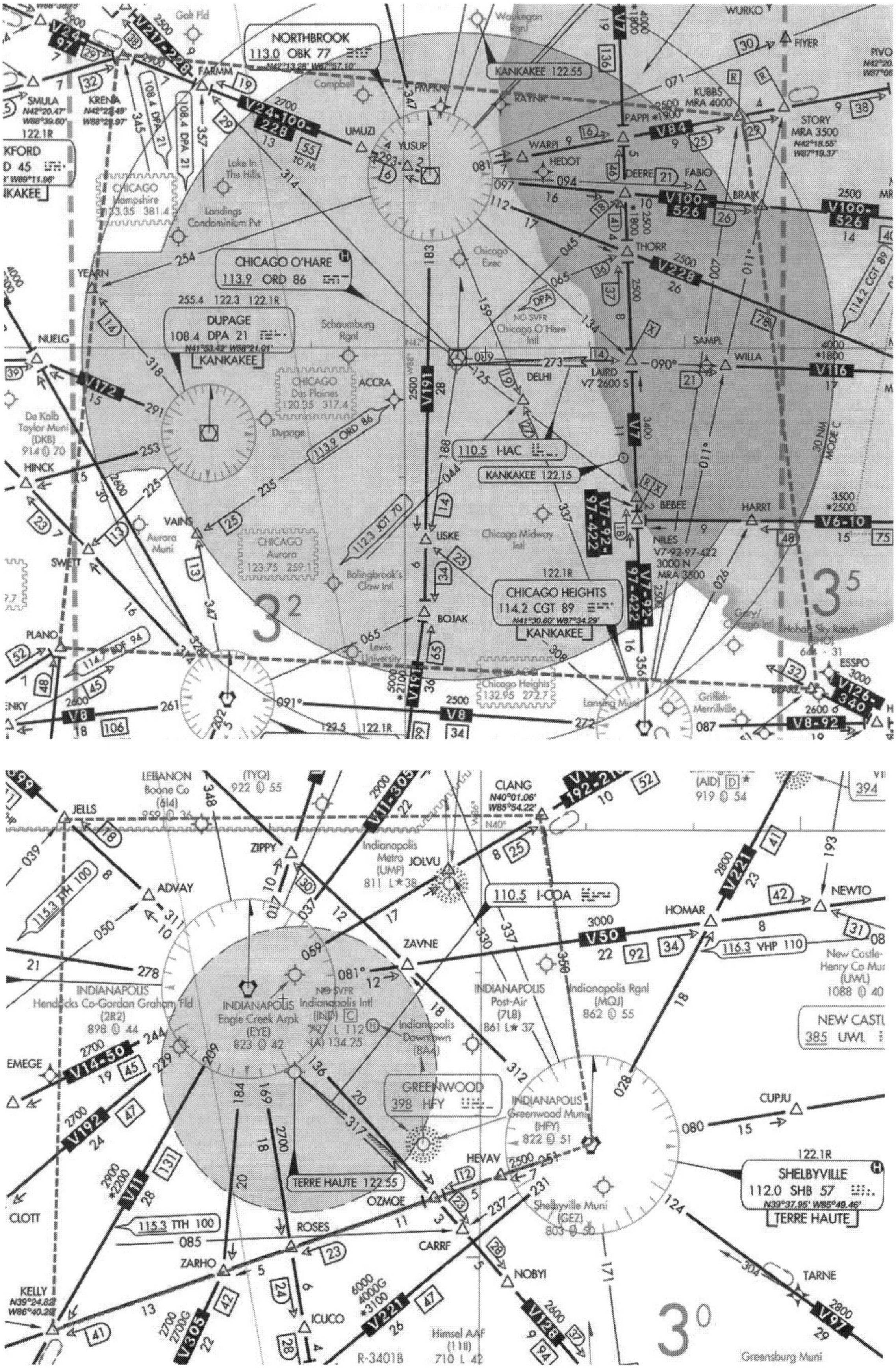

Figure 5-5. Box and cornerpost system of organization.

At airports where box and cornerpost systems are used, the LOA usually specifies that inbound IFR aircraft enter the approach control's airspace at one of the cornerposts; thus, these areas are known as arrival gates. The LOA will then also stipulate that departures remain clear of the cornerposts by departing through the "sides" of the box (known as departure gates). Based on traffic conditions and runways in use, the approach control facility advises adjacent ARTCCs of the spacing that must be maintained as aircraft fly inbound to the airport. This is usually defined in terms of "miles-in-trail". Typically, an approach control would require either 5, 10, or 15 miles in trail over any particular cornerpost fix. It is the ARTCCs job to properly sequence these inbound aircraft. Normally that process starts hundreds of miles out, with inbound aircraft properly spaced and sequenced, crossing "metering" fixes as they converge closer to the arrival airport. (Figure 5-6).

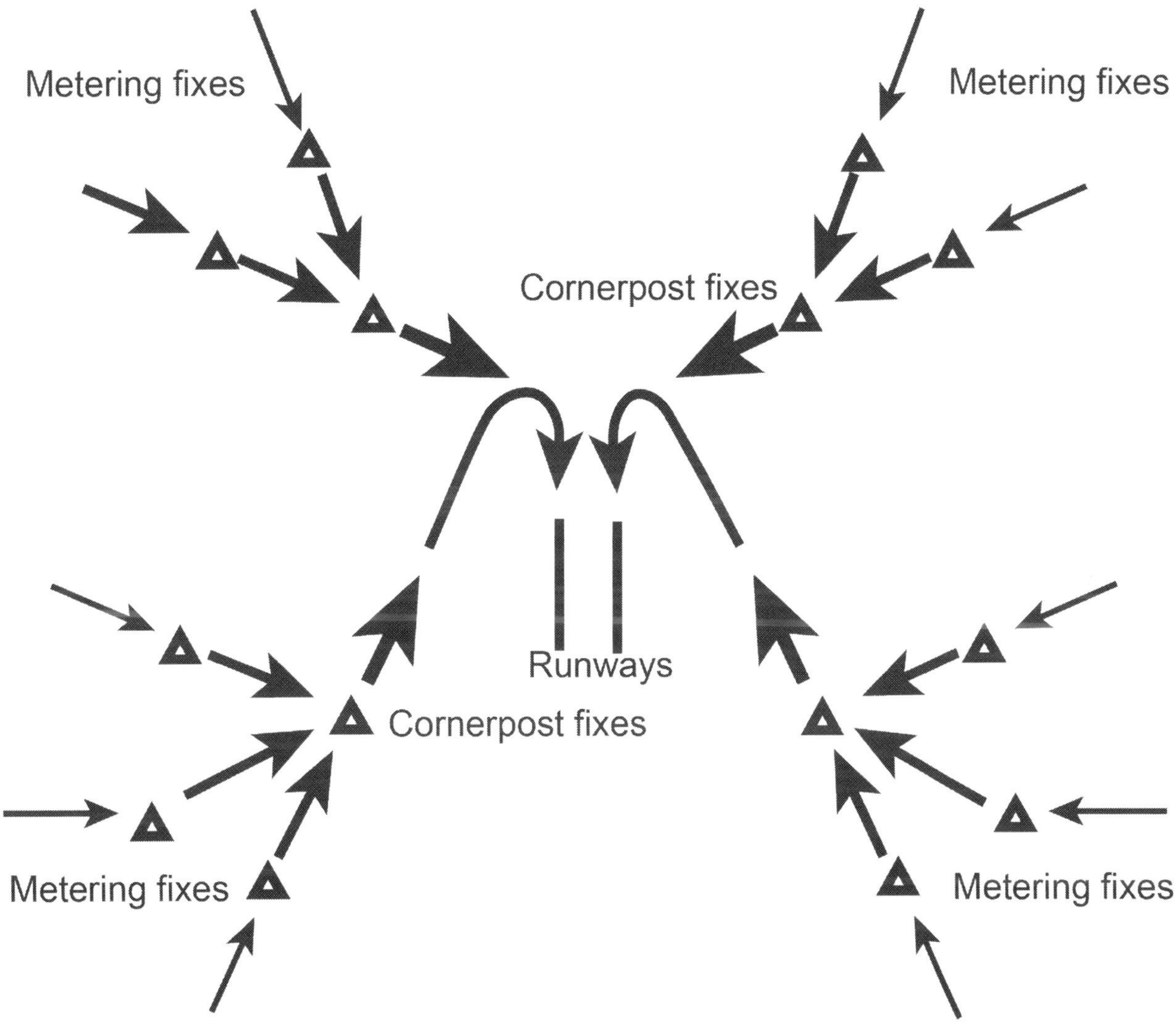

Figure 5-6. Cornerpost and metering fix traffic flow.

Approach Control Operations

The ARTCC controller insures that the aircraft is at (or descending to) the altitude specified in the letter of agreement and advises the pilot to contact approach control when approaching one of the arrival fixes. Depending on the airport configuration and the amount of traffic being handled, up to a dozen different controllers might be assigned approach control responsibilities. Generally, there is at least one sector for each inbound "corner", a couple of final controllers who merge these aircraft "streams" onto the approach to the runway, and possibly a final approach monitor controller. In addition, there are controllers assigned to separating departing aircraft, overflights and operations at nearby satellite airports.

Generally, arrival controllers will be delegated airspace above about 6,000' agl or so and must keep their aircraft within a narrow corridor while descending. This corridor is designed to keep inbounds separated from departing aircraft while giving the controllers room to slow down, descend and line up the aircraft. Depending on the runways in use, the arrival controllers essentially establish all the aircraft on a wide and long traffic pattern consisting of a downwind (either left or right) or a base leg. The final controllers then merge, sequence and space these streams of aircraft and clear them for the appropriate instrument approach.

Arrival Controller Responsibilities

The process described herein has been simplified; in reality it is much more complicated with many details, but this description should provide the general concept of the operation of a TRACON. The following examples generally describe the flow of inbound aircraft to the Indianapolis airport. Normally one runway (5L) is used for arrivals while the other parallel runway (5R) is used for departures (Figure 5-7).

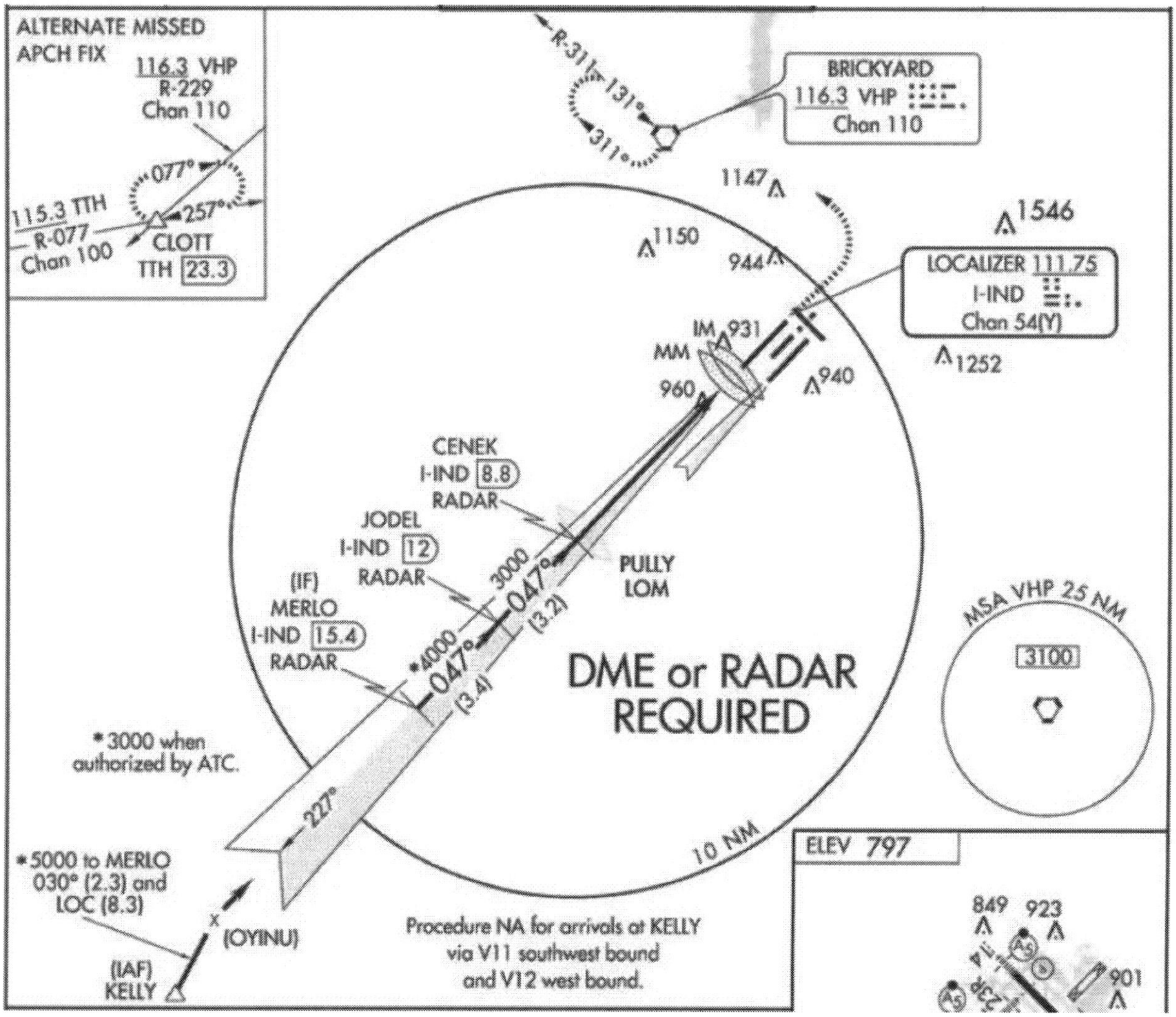

Figure 5-7. Plan view of ILS runway 5L approach to Indianapolis.

Typically, the ARTCC brings each inbound aircraft over one of the cornerposts either level at or descending to 11,000 feet. Once the aircraft has entered Indianapolis Tracon's assigned airspace, the arrival controller is permitted to descend the aircraft to 10,000 feet. Every aircraft inbound to Indianapolis is then vectored toward the airport and sequenced behind other inbound aircraft. The arrival controller is authorized to descend the aircraft if coordination has been accomplished with the appropriate departure controller.

If the aircraft is arriving straight in from the southwest (Kelly), once the aircraft is within about 15 miles of the airport, the arrival controller descends the aircraft for the ILS approach. If the aircraft is inbound from any other direction, once it is within about 15 miles of the airport, the controller will descend the aircraft to 3,000' and place it on either a left or right extended downwind leg. When the aircraft is abeam the airport, the pilot is advised to contact the final approach controller, who is tasked with merging the left and right downwind traffic with the straight-in traffic arriving from the southwest. The final controller determines the proper approach sequence and spaces the aircraft using either vectors or speed restrictions or a combination. When an aircraft is in the proper position, adequately separated from both preceding and following aircraft, the controller vectors the aircraft to intercept the runway localizer to complete the approach and advises the pilot to contact the local controller in the tower.

When vectoring an aircraft onto the final approach course, the controller must ensure that it is separated at all times. This task can be fairly difficult since the controller is usually required to sequence aircraft with different flight characteristics to the same final. Even two identical aircraft may not fly the approach at the same speed because of aircraft loading characteristics, pilot preferences, or any number of other variables unknown to the controller.

Besides ensuring separation, the arrival controller must also ensure that the aircraft is positioned such that the pilot can make a safe and gradual transition to the final approach course. Since the controller has assumed navigational responsibility, each aircraft must be vectored into the proper position and at an appropriate heading to ensure that the pilot can safely transition to the final approach course. To facilitate this transition, the FAA handbook specifies criteria that the controller must maintain during this procedure.

The handbook specifies that an approach gate exists along every final approach course whenever radar vectoring to that instrument approach is in progress (Figure 5-8). The approach gate is located either 1 nm outside the final approach fix or 5 nm from the end of the runway, whichever distance is greater. Every aircraft vectored to the final approach course must intercept the final approach course no less than 2 nm outside the approach gate. This requirement can be relaxed only if requested by the pilot, but in no case may the aircraft be permitted to intercept the final approach course inside the final approach fix.

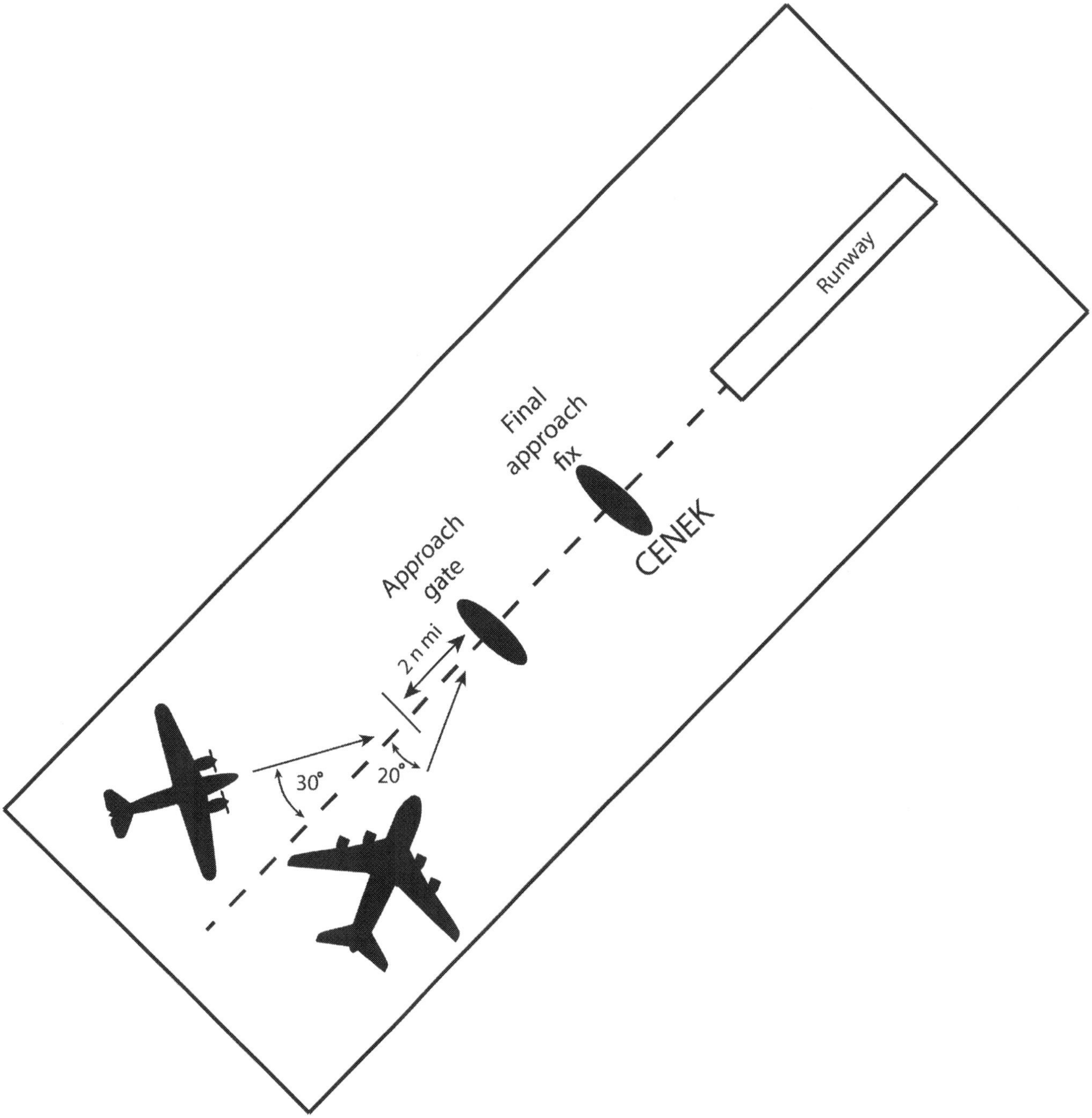

Figure 5-8. Final approach intercept and the approach gate.

The controller must also ensure that the aircraft intercepts the final approach course at a sufficiently shallow angle to permit a smooth transition to the final approach course. The FAA handbook specifies that if the aircraft intercepts the final approach course at a point less than 2 miles from the approach gate, the intercept angle should be less than or equal to 20°. If the aircraft will intercept the final approach course 2 miles or farther from the approach gate, the aircraft may intercept the final approach course at an angle no greater than 30°. Figure 5-9 is a radar display of the final approach courses for runways 5L and 5R with the outer markers and other fixes delineated.

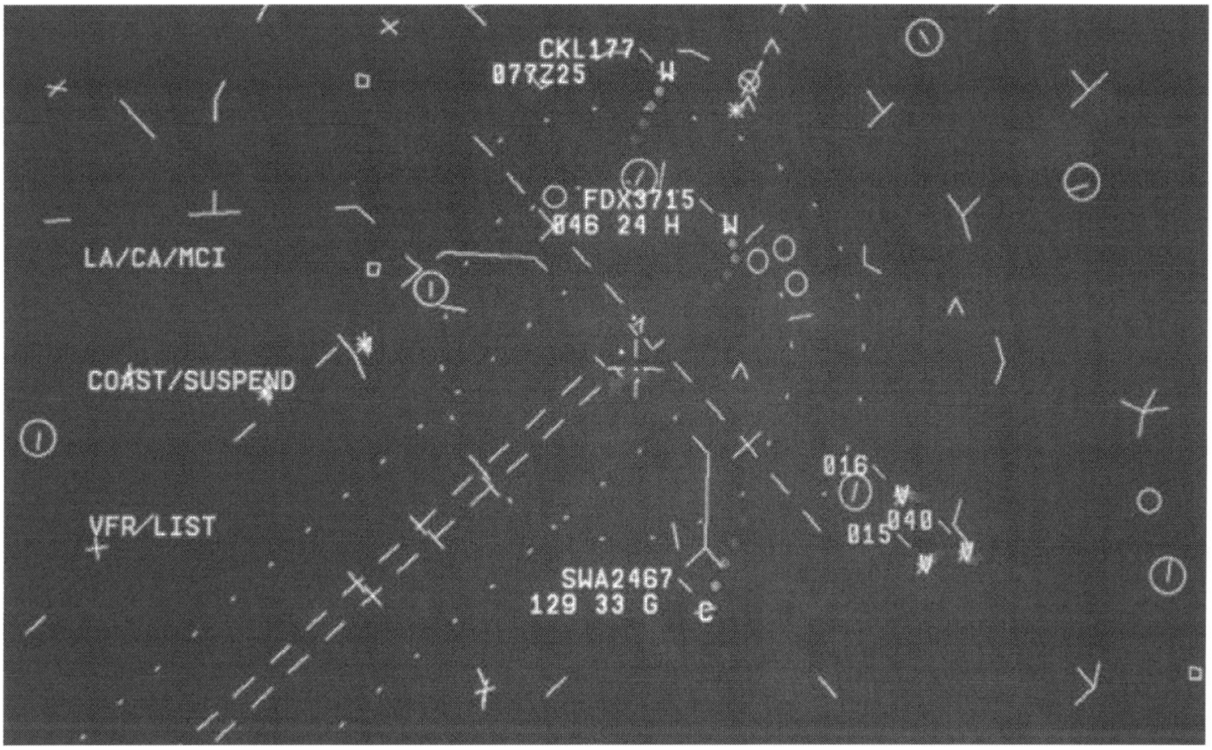

Figure 5-9. Indianapolis radar display for runways 5L and 5R.

Approach Clearance

The following information must be issued to the pilot before the aircraft intercepts the final approach course:

- The aircraft's position relative to a fix associated with the instrument approach. This fix is usually the final approach fix but may be any other navaid or intersection along the final approach course.
- A vector that will cause the aircraft to intercept the final approach course at the proper point and at an allowable intercept angle.
- The actual clearance to conduct the instrument approach.
    - If the aircraft is not on a published transition route to the final approach course, the controller must assign the aircraft an altitude that is no lower than the minimum vectoring altitude. The aircraft must remain at or above this altitude until it is established on a published segment of the instrument approach.
- The controller should issue instructions to contact the next controller, if appropriate, and the frequency to be used.

Here is an example of phraseology that would be used when vectoring an aircraft to the ILS runway 5L approach.

> "Brickyard three eleven, seven miles from CENEK, turn right heading zero two zero, intercept the final approach course at or above three thousand, cleared for the ILS runway five left approach. Monitor Indianapolis tower on one two zero point niner, report the outer marker inbound."

Minimum Vectoring Altitude

While operating along a federal airway, an approved transition route or during the conduct of an instrument approach, the pilot must comply with the minimum altitudes provided on appropriate navigation charts. Once an aircraft has been vectored off one of these published routes, however, it becomes the controller's responsibility to ensure that the aircraft remains safely above terrain or local obstructions. To assist in this task, the FAA has developed minimum altitudes and has provided these altitudes to controllers at every radar-equipped facility (see Figure 5-10). In the TRACON environment these altitudes are known as minimum vectoring altitudes (MVA); in the enroute environment they are called minimum IFR altitudes (MIA). The use of these altitudes is mandatory and provides each aircraft with standard IFR separation from any terrain or obstacle. In general, aircraft operating at minimum vectoring altitudes must remain at least 3 nautical miles laterally from or at least 1,000 feet above any obstruction or terrain in non-mountainous areas, and 2,000 feet in mountainous areas.

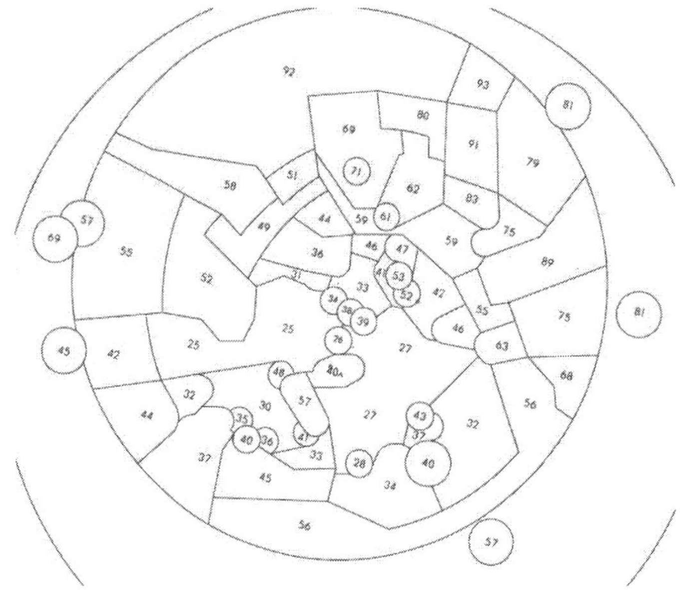

Figure 5-10. Typical MVA chart for use at Phoenix TRACON.

Departure Control

Depending on the complexity of the facility, departure control may be operated by the approach controller, a separate control position, or could even be divided into a number of different subsectors. In any case, it is the responsibility of the departure controller to separate departing aircraft while still complying with appropriate facility procedures. Once the aircraft has taken off it is changed to the departure controller's frequency. The departure radar controller must radar identify the aircraft and verify the accuracy of the aircraft's mode C transponder. The easiest identification method used is to observe a departing aircraft target within 1 mile of the takeoff runway end. The aircraft's mode C altimeter function must also be verified. This is typically accomplished by the pilot stating his or her altitude on initial contact. (Figure 5-11).

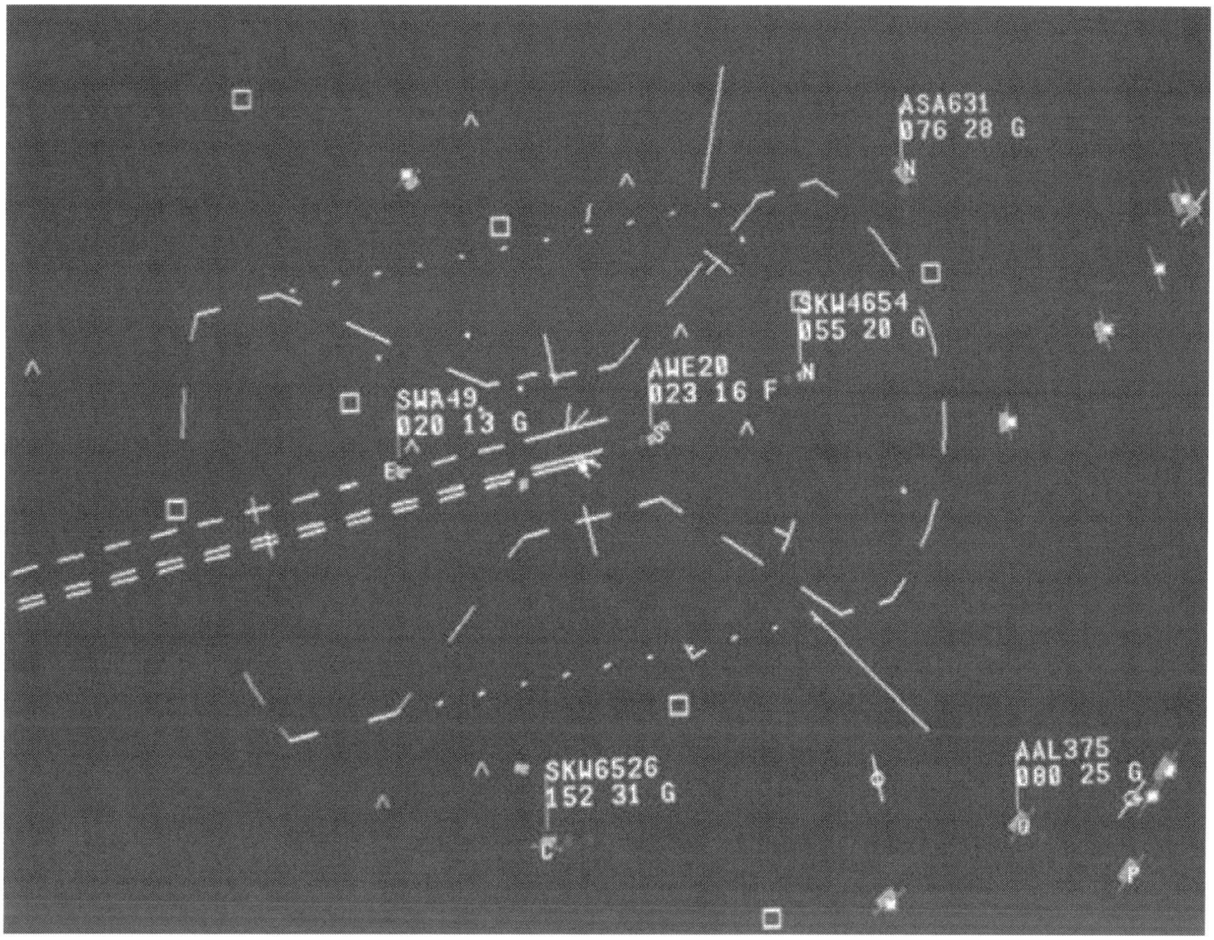

Figure 5-11. Phoenix departure control radar display.

After radar-identifying the aircraft, the controller will usually assign a higher altitude, generally the top of the TRACON airspace. The departure controller must insure that the aircraft stays within both the vertical and horizontal limits of the controller's assigned airspace. (Figure 5-12). If the aircraft needs to be turned on course, the controller will do so when able. Once radar contact has been established and the pilot has been advised, the controllers are permitted to use radar separation. However, they are not prohibited from using nonradar separation if that provides an operational advantage.

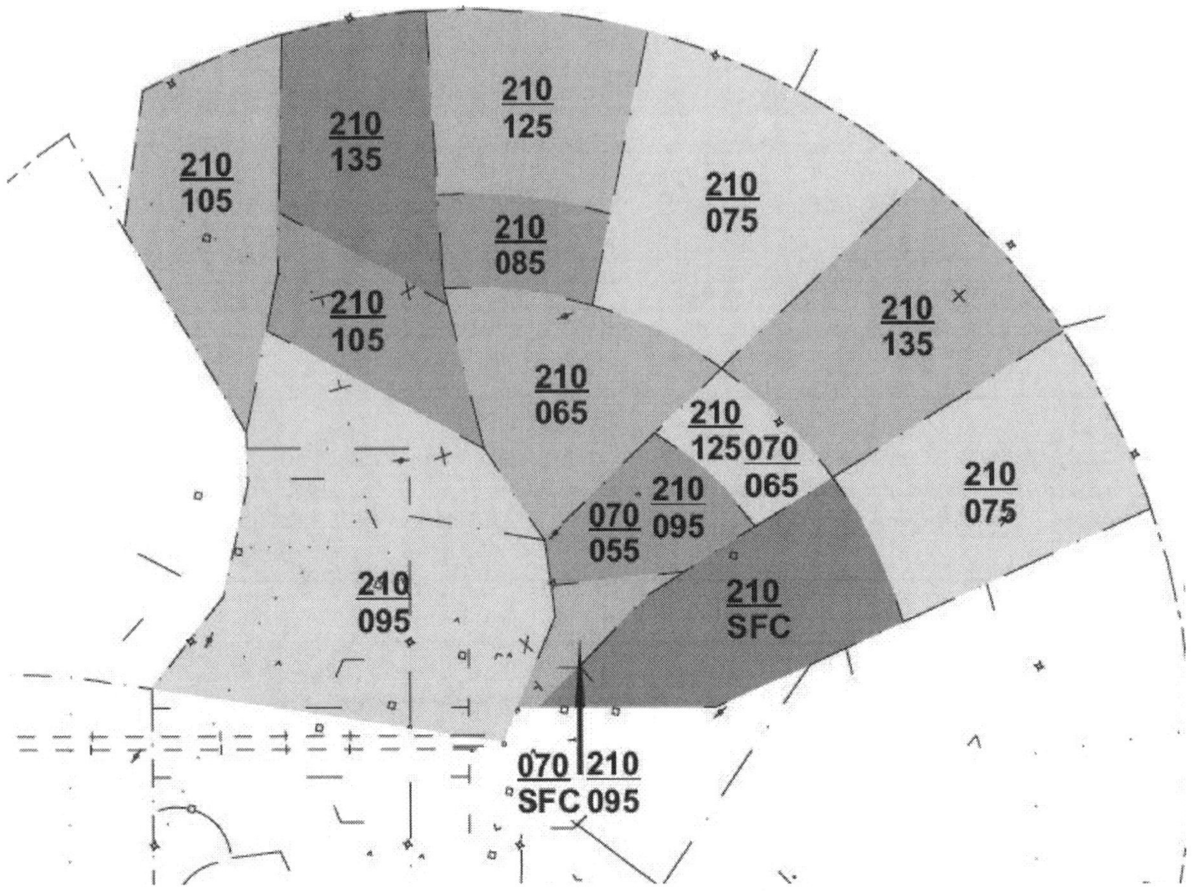

Figure 5-12. Airspace allocation for Phoenix TRACON north and east departure controller.

The controller may vector the aircraft to join the route of flight while still complying with LOAs and facility directives. The controller will attempt to clear the aircraft to climb to the pilot's requested altitude as soon as is practical. If this is not possible because of a lack of jurisdiction or traffic conflicts, the aircraft will typically be cleared to the altitude closest to that filed by the pilot.

If the aircraft will transit other subsectors within the terminal facility, the departure controller must either hand off or point out the aircraft to the appropriate controllers. Such handoffs are accomplished manually or through the use of automated procedures. If the aircraft is remaining at a fairly low altitude, it will usually be handed off to an adjoining terminal facility. But, if the aircraft will fly at a relatively higher altitude, it is generally handed off to the appropriate ARTCC who will climb the aircraft to their cruising altitude.

Overflights and Satellite Airports

At most TRACONs, there are smaller airports (known as satellite airports) that will occasionally have IFR arrivals and/or departures. There are usually additional control sectors assigned to handle these aircraft. There is also a need to separate aircraft that might be transiting through the TRACON's airspace. Procedures have been developed to handle each of these types of operations but are usually unique to each facility. But in general, these sectors are granted use of a limited amount of airspace to separate these aircraft from the routes heavily used by inbound and outbound aircraft at the primary airport.

Radar Traffic Information

Controllers can also use radar to assist pilots in avoiding potential conflicts with other aircraft by advising them of the relative position and altitude of any potentially conflicting traffic. This is known as providing traffic advisories. A controller may provide traffic advisories to any aircraft, whether IFR or VFR, on a workload-permitting basis. Priority is still given to providing separation to aircraft that are participating in the ATC system.

Controllers phrase the advisory in the following manner:

- The azimuth of the conflicting traffic in terms of a 12-hour clock relative to the aircraft's ground track.
- The aircraft distance in nautical miles.
- The direction in which the conflicting traffic is proceeding.
- The type of aircraft, if known, and the altitude.

The following are sample traffic advisories.

> "FedEx two eleven, traffic twelve o'clock, three miles, eastbound, type and altitude unknown."
>
> "American eleven sixteen, traffic, one o'clock, two miles, westbound Cessna 150, one two thousand."
>
> "Brickyard nineteen fifty two, traffic nine o'clock, five miles, southbound, a VFR military trainer, last reported altitude six thousand."

Traffic advisories should be provided early enough so that the pilot has ample time to locate the other aircraft or to ask the controller for a traffic avoidance vector. If the pilot requests such a vector, the controller should issue a heading that will prevent the targets from merging. If the controller is unable to do so, because of other conflicting traffic or procedural restrictions, the pilot should be informed of the reason.

Conflict Alert

Many of the computerized beacon processing systems used by the FAA are capable of projecting the flight paths of aircraft and can alert the controller in advance of certain potentially unsafe conditions. Once an alert has been sounded by the radar system, it is the controller's responsibility to resolve the situation if two IFR aircraft are involved, or to advise the pilot to resolve the situation if a VFR aircraft is involved. One of the most important of these safety systems is the Conflict Alert software program. Conflict alert uses a tracking program to predict when two tracked aircraft will approach each other with less than the required vertical, lateral, or longitudinal separation minima. If conflict alert predicts that this might occur, the computer alerts the controller, who can then evaluate the situation and initiate corrective action if needed.

In certain terminal areas, however, conflict alert constantly predicts hazardous situations that do not really exist. Since the computer software is unaware of the controller's or the pilot's planned actions, it can only predict an aircraft's future ground track based on its history. This may cause false alerts to be routinely sounded at a busy approach control facility where the airspace configuration is necessarily complex. For example, at a busy airport where aircraft are being vectored for parallel runways, there is usually a point when both aircraft are in a position where, if they are not turned, they are likely to conflict. This situation is perfectly acceptable since the controller is planning to turn each aircraft toward the appropriate runway. But because conflict alert has no way of knowing this, it sounds a warning each time two aircraft are in this position. This routine sounding of an alert is distracting and potentially dangerous, because every time the alert is sounded, the controller must verify whether an actual problem exists, diverting attention away from other traffic. In addition, the routine sounding of a misleading alert will eventually cause controllers to disregard any alarm produced by the system. For these reasons, FAA computer programmers inhibit conflict alert in areas where numerous false alarms are commonly generated.

Minimum Safe Altitude Warning

In the past, there have been accidents caused when pilots inadvertently flew their aircraft too low and hit either terrain or man-made obstacles. To assist air traffic controllers to detect aircraft that are at or approaching an altitude that may be in close proximity to the ground or to obstructions, the FAA has implemented a computer software program known as **minimum safe altitude warning (MSAW).** MSAW uses the mode C altitude encoder on the aircraft and the radar computer system tracking capabilities to predict whenever a tracked aircraft is in imminent danger of colliding with an obstacle and warns the controller. Because aircraft conducting instrument approaches must necessarily descend closer to the ground than the MSAW system permits, allowances must be made in these areas. The MSAW software relaxes the obstacle avoidance criteria but still monitors aircraft between the final approach fix and point 2 nautical miles from the approach end of the runway.

MSAW is designed to predict both unreasonably low altitudes and excessive aircraft descent rates that might prove to be dangerous. Every aircraft within the approach area is monitored by the radar system, and an alert is sounded if an aircraft descends 100 feet below the minimum altitude for that segment of the approach. In addition, the computer uses past altitude information to extrapolate the aircraft's current rate of descent. If it determines that the aircraft is currently above the minimum altitude for that segment but is predicted to descend 200 feet below the minimum altitude within the next 15 seconds, the controller is alerted. Because of differing aircraft types and approach minima for each runway, the MSAW software is inhibited within 2 nautical miles of the approach end of the runway.

Whenever an unsafe condition is predicted by the MSAW software, an alert is sounded and the letters "Low Alt" begin to flash in the aircraft's data block. When this occurs, the controller must immediately evaluate the situation and, if appropriate, issue the pilot a verbal warning.

"Low altitude alert, Cessna four two two papa alpha, check your altitude immediately, the MVA in your area is three thousand two hundred, altimeter two niner eight six."

It is then up to the pilot to evaluate the situation and determine what actions are necessary to return the aircraft to the proper flight path.

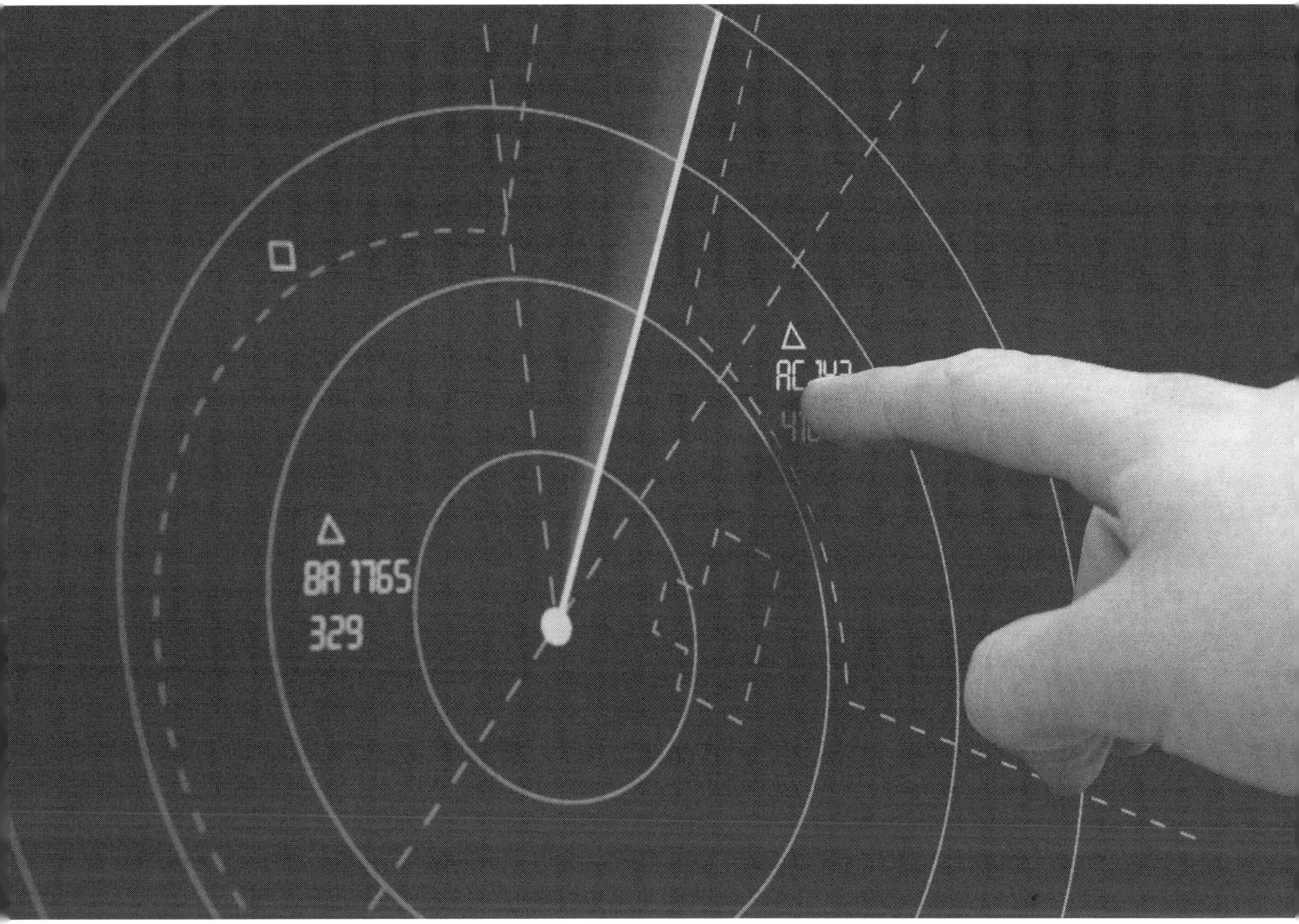

Chapter 6

Enroute Air Traffic Control

Although control towers may be one of the most publicly visible components of the air traffic control system, much of the complexity and operation of the overall ATC system is managed by the ARTCCs. As previously described, TRACONs and control towers are responsible for much of the air traffic control within the immediate vicinity of busy airports. Most of the decisions made by controllers at those facilities have an immediate impact on the affected aircraft and traffic flow, but generally do not affect the entire ATC system. As such, you might say that those facilities are more "tactical" in nature than ARTCCs.

Although the primary function of an ARTCC is to separate aircraft traveling between airports, an equally important role is the organizing of overall traffic flows across the country. The ARTCC and its controllers manage this "strategic" flow of aircraft across the country enabling a more effective and efficient use of airspace. Consequently, traffic flow management (TFM) is a much larger part of an ARTCC controller's job than it is at other ATC facilities.

ARTCC Organization

The FAA has established the areas of authority for every ARTCC. Although seldom used in the United States, ICAO uses the name Area Control Centre (ACC) to describe what the US calls an ARTCC. The proper term for the airspace controlled by an ARTCC (or ACC) is a Flight Information Region (FIR). Therefore, the airspace allocated to Chicago ARTCC is more properly called the "Chicago Flight Information Region". As the lead ATC facility in the FIR, Chicago ARTCC has information on all IFR flights in that area. Finally, although they are officially known as an ARTCC, many participants in the NAS refer to this facility simply as "center."

FAA documentation specifies the physical boundaries of each ARTCCs airspace. Furthermore, each ARTCC has delegated certain areas for use by towers or TRACONs; LOAs have been established between these facilities that document boundaries, as well as the procedures to be used when aircraft cross airspace boundaries.

Figure 6-1. ARTCC boundaries.

Since ARTCC's use radar and communication equipment that can operate off-site, they are not always located within the same city that bears their name. The 22 ARTCCs in operation across the U.S. are located in the following cities:

TABLE 6-1 ARTCC LOCATIONS

ARTCC Facility	Location
Albuquerque	Albuquerque, New Mexico
Anchorage	Anchorage, Alaska
Atlanta	Hampton, Georgia
Boston	Nashua, New Hampshire
Chicago	Aurora, Illinois
Cleveland	Oberlin, Ohio
Denver	Longmont, Colorado
Fort Worth	Euless, Texas
Honolulu	Honolulu, Hawaii
Houston	Houston, Texas
Indianapolis	Indianapolis, Indiana
Jacksonville	Hilliard, Florida
Kansas City	Olathe, Kansas
Los Angeles	Palmdale, California
Memphis	Memphis, Tennessee
Miami	Miami, Florida
Minneapolis	Farmington, Minnesota
New York	Ronkonkoma, New York
Oakland	Fremont, California
Salt Lake City	Salt Lake City, Utah
Seattle	Auburn, Washington
Washington	Leesburg, Virginia

Domestic Enroute Air Traffic Control

Every ARTCC's airspace is divided into numerous smaller areas called sectors. Each of these sectors is fashioned in a logical manner, taking into consideration airway structure and traffic flows. Sectors are designed to have a workable amount of traffic for a team of controllers on a typical day, and to make it easier for an individual controller to separate and organize the traffic flow within their sector. Every ARTCC's airspace is partitioned into twenty or more sectors. Figures 6-2-1 and 6-2-2 diagram the low and high-altitude sectors of ARTCC airspace.

Figure 6-2-1. Low altitude ARTCC sectors.

Figure 6-2-2. High altitude ARTCC sectors.

The airspace at most centers is stratified into at least two vertical levels: a low-altitude group of sectors extending from the Earth's surface up to 23,000 msl (FL 230), and a group of high-altitude sectors extending upwards from 24,000 msl (FL 240). Busier centers may stratify into three levels, in which the low-altitude sectors extend from the ground to FL230, high-altitude sectors from FL 240 to somewhere around FL 330-350, and then ultra-high sectors, which lie above the high-altitude sectors. The ultra-high sectors (if used) usually conform to the same lateral dimensions and layout as the high-altitude sector structure, although they can encompass more than one lower sector, as the volume of traffic is not nearly as high in these areas.

The physical dimensions of each sector within an ARTCC are specified in facility directives known as Standard Operating Procedures (SOP) guide. SOPs are similar to LOAs but apply only to controllers working within a particular facility. Among many other things, SOPs specify the horizontal and vertical boundaries of each sector and describe the procedures to be used when aircraft cross the boundary between sectors.

Traffic Flows Within an ARTCC

There are generally three types of sectors within an ARTCC. The first is primarily tasked with organizing inbound traffic flows to busy airports. The second is more concerned with departing aircraft. The third type of sector handles primarily enroute traffic. Figure 6-3 shows the traffic inbound and outbound from the Chicago O'Hare airport. You can easily see the sector organization around those traffic flows. Each sector has been physically and operationally designed around the primary traffic flow found in that sector. But elements of each type of operation can be found in every ARTCC sector. Figure 6-4 adds non-O'Hare traffic to the mix. However, since most of the overflying traffic is at a much higher altitude, it does not affect traffic arriving or departing at O'Hare as much as it appears. As you can see, this additional traffic makes it much harder to recognize the flows into and out of O'Hare. However, most ARTCCs have equipment that allows controllers to monitor only the low altitude traffic if they are only responsible for that traffic, making it easier to identify the aircraft controllers need to separate and sequence.

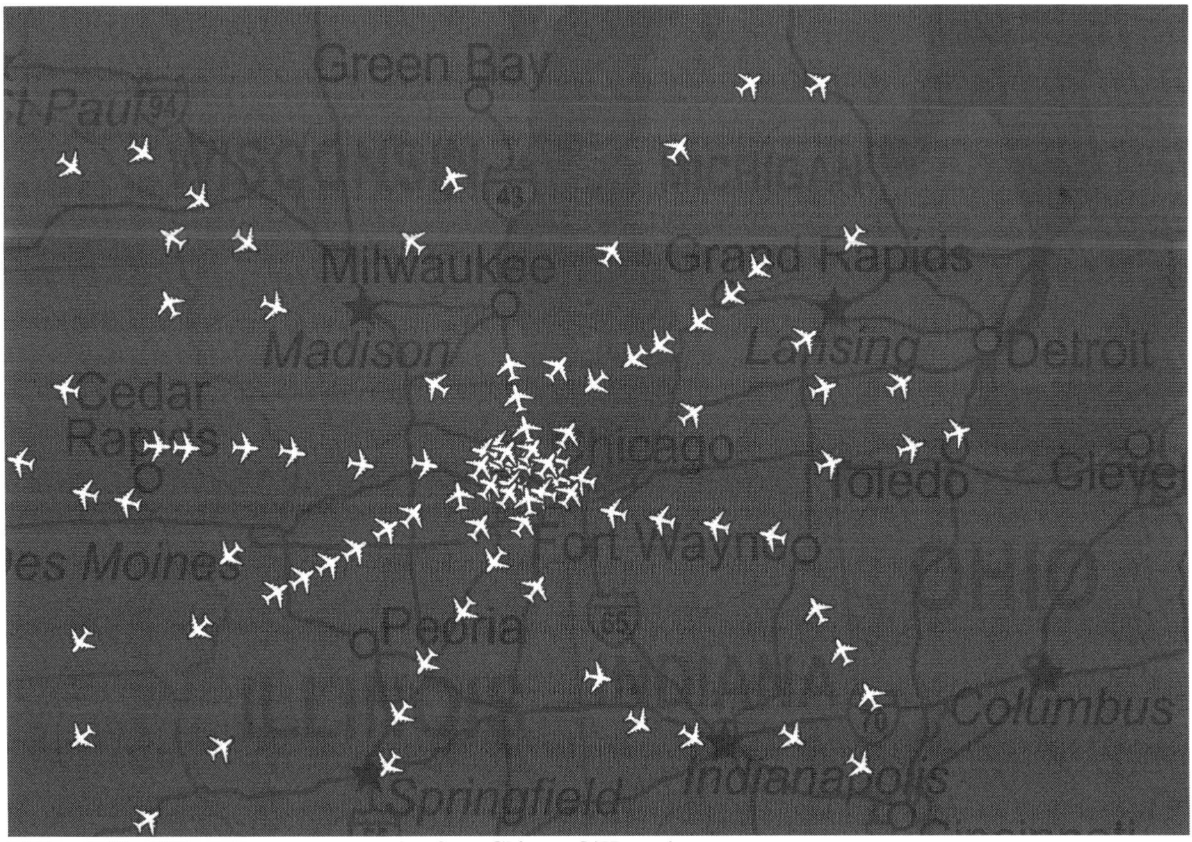

Figure 6-3. Traffic arriving and departing from Chicago O'Hare airport.

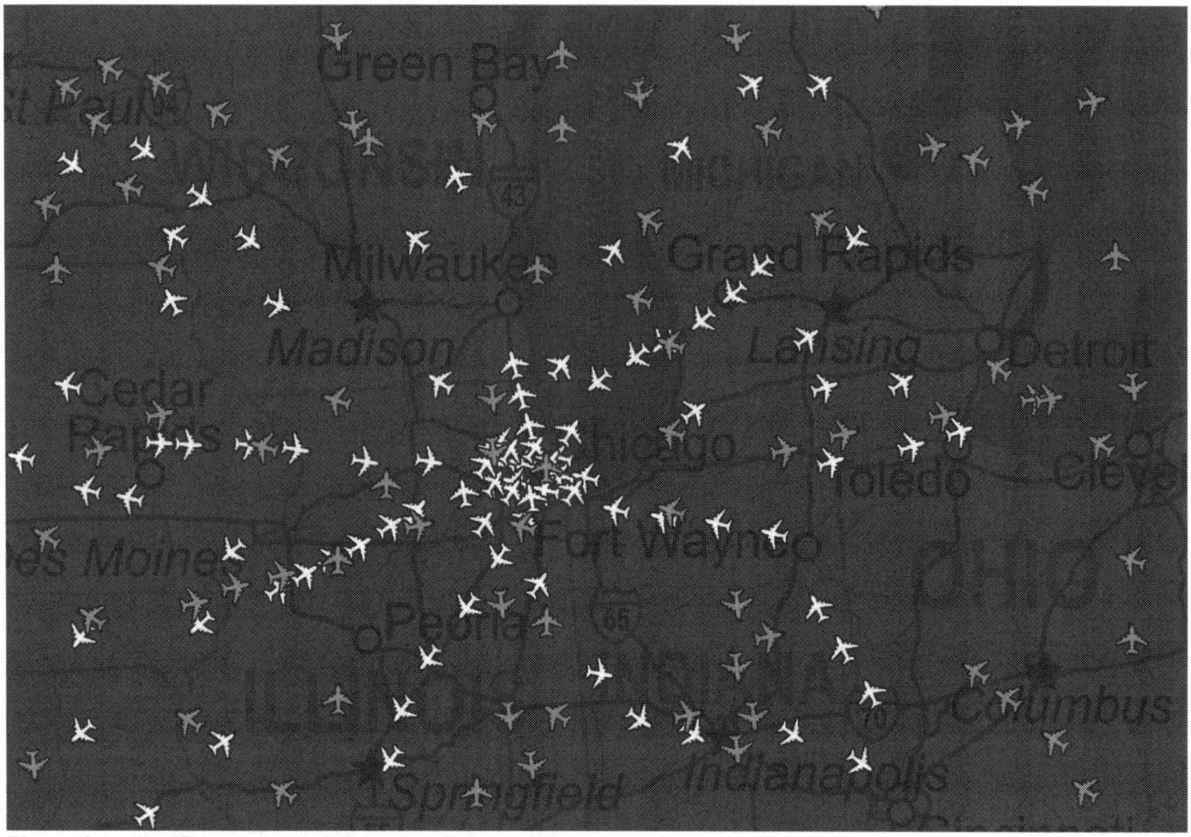

Figure 6-4. Traffic arriving and departing Chicago O'Hare as well as overflying through the airspace of Chicago Center.

Preferred Routes

To better organize ARTCC enroute traffic flows, the FAA has developed a system of designated preferred routes and altitudes for flights between major airports. Some of these routes are published, while others are internally assigned by the FAA. The use of preferential routes and altitudes enhances traffic flows, thereby reducing controller workload and increasing overall system efficiency. Preferred routes vary depending upon traffic flow, day of the week and time of day, aircraft type, speed, altitude and navigation capability as well as forecast or actual enroute traffic restrictions. There are over 10,000 preferred routes in the FAA database.

When more than one airway extends from one airport to another, it is common practice to designate each as a one-way airway. This procedure reduces the chance of a head-on collision at or near a sector or facility boundary, as aircraft departing an airport will all be assigned a certain airway or route for the climb to altitude, and inbound aircraft will be assigned a different, non-intersecting route for their descent. If there are insufficient airways to designate one-way airways between facilities, specific altitudes are usually reserved for inbound aircraft, while other altitudes are used by outbound aircraft. A sampling of preferred routes departing O'Hare airport follows:

TABLE 6-2 SAMPLE FAA PREFERRED ROUTES OUT OF CHICAGO'S O'HARE AIRPORT

Origin	Destination	Route
ORD	ATL	ORD CMSKY CARYN CYBIL PXV J73 BNA RMG5 ATL
ORD	ATL	ORD CMSKY CARYN CYBIL PXV BNA KOLTT1 ATL
ORD	BOS	ORD ELX CRL J554 JHW J82 ALB GDM GDM4 BOS
ORD	BOS	ORD EBAKE WISMO POSTS PADDE SVM DKK ALB GDM4 BOS
ORD	DCA	ORD GIJ J146 J34 BUCKO FRDMM2 DCA

Origin	Destination	Route
ORD	DCA	ORD GIJ J146 J34 BUCKO NUMMY2 DCA
ORD	DFW	ORD RBS STL RZC FSM RRNET SEEVR1 DFW
ORD	DFW	ORD RBS STL RZC FSM RRNET BRDJE1 DFW
ORD	DFW	ORD RBS STL RZC FSM WILBR4 DFW
ORD	DFW	ORD ACITO ADELL ARLYN STL RZC FSM RRNET SEEVR1 DFW
ORD	DFW	ORD ACITO ADELL ARLYN STL RZC FSM RRNET BRDJE1 DFW
ORD	DFW	ORD ACITO ADELL ARLYN STL RZC FSM WILBR4 DFW
ORD	EWR	ORD ELX CRL J584 SLT FQM3 EWR
ORD	EWR	ORD DUFEE ELX HAAKK DOXXY SOSIC KEEHO J584 SLT FQM FQM3 EWR
ORD	IAD	ORD MOBLE ADIME OTENS ANEWA RIEKE DJB J34 AIR MGW GIBBZ2 IAD
ORD	IND	ORD EON V24 VHP IND
ORD	IND	ORD EARND ELANR EMMLY JAKKS IND
ORD	JFK	ORD ELX CRL J554 JHW J70 LVZ LENDY6 JFK
ORD	JFK	ORD DUFEE ELX HAAKK DOXXY SOSIC JHW J70 LVZ LENDY6 JFK
ORD	LGA	ORD GIJ J146 MIP MIP4 LGA
ORD	LGA	ORD MOBLE ADIME GERBS J146 ETG MIP4 LGA

Fixes where aircraft cross facility boundaries into or out of TRACON airspace are known as gates. LOAs are very specific about the procedures, altitudes, and airways to be used at each gate. Controllers can sometimes use different altitudes and crossing points if it has been approved (this is known as coordination, an approval request or APREQ). SOPs are just as specific, defining the routes and altitudes that should be used by aircraft crossing sector boundaries within the facility. As in facility crossings, any procedure can be modified with an APREQ if both controllers agree.

Controller Duties in an Air Route Traffic Control Center

There are many different job duties in an ARTCC, but in general, three basic working positions monitor and direct traffic. These include radar, radar associate, and traffic management unit controllers. There are also a number of management and supervisory positions as well as training, quality control, and airspace specialists. Additionally, ARTCC's have individuals responsible for maintaining equipment, known as airway facilities (AF) and system software, known as automation. There are numerous other employees in the building as well. Most ARTCCs employ several hundred controllers, as well as other employees involved in supporting the overall mission. A list of positions within ARTCC follows:

TABLE 6-3 ARTCC POSITIONS

Designator	Position
A	Developmental Controller
C	Coordinator
D	Sector Controller
DR	Radio Controller
DSC	Data Systems Coordinator
ERM	ERM Route Metering
FDCS	Flight Data Communications Specialist
M	AMIS Controller
MC	Mission Coordinator
OM	Operations Manager
OS	Operations Supervisor

Designator	Position
R	Radar Controller
RA	Radar Associate
SDCS	Supervisory Data Communications Specialist
STMCIC	Supervisory Traffic Management Coordinator–in–Charge
TMC	Traffic Management Coordinator
WC	Weather Coordinator

Radar Controller

Every sector is staffed by a controller whose primary responsibility is to separate aircraft using radar. Radar controllers issue altitudes, headings, or airspeed changes that keep aircraft separated and in compliance with LOA's and SOP's. This controller is generally the person ultimately responsible for separating aircraft within the sector; they are also the person who communicates directly with aircraft. Coordination with other sectors and facilities is a duty shared by the radar and the associate controller. Finally, the radar controller is responsible for ensuring handoffs are performed in a timely manner. This position is ordinarily known as the R-side.

Radar Associate/Non-radar Controller

In times of light traffic, the sector may be operated by a single controller accomplishing all the above-mentioned duties. But if traffic warrants, a sector may also be staffed by a radar associate/nonradar controller whose duties are to assist the radar controller. The associate controller's duties include ensuring flight data information accurately reflects the aircraft's position, altitude, and route of flight. The associate controller works with the radar controller to plan separation of aircraft and to coordinate with other sectors and facilities. They are also responsible for delivering IFR clearances to airports not served by a TRACON and providing nonradar separation to aircraft not covered by the sector's radar capabilities. As the name implies, they are also responsible for nonradar separation when needed. This position is commonly known as the D-side.

Traffic Management Unit coordinators

Traffic management units (TMU) are established at every ARTCC and many high-density terminal facilities. It is the traffic management coordinator's (TMC) responsibility to balance air traffic demand with the capacities of the national airspace system. The TMC coordinates traffic flows within and between ARTCCS and implements real time flow patterns to maximize ATC efficiency. The TMC must maintain a continuous awareness of both local and national traffic flows, weather conditions and traffic forecasts as well as any existing enroute and/or terminal delays. They must also pay attention to airport delays at high-density airports, such as Newark or Dallas. As conditions in the NAS change, TMCs issue instructions that adjust traffic flows to either manage aircraft within areas of traffic constraints or avoid them completely when possible.

Flight Data and Radar Data Processing

Controllers in an ARTCC work primarily with two different kinds of data; radar data and flight plan information. Radar data is managed using a system called radar data processing or RDP. The RDP system combines radar information obtained from multiple radar sites spread across the country. Enroute radars use the radar return from several different radar sites and coalesce them into one radar return; this is known as a mosaic presentation. This data is digitized, analyzed, and then routed to each controller for display. The flight data processing (FDP) system provides controllers with information concerning each aircraft, such as altitude and routing, as well as pertinent airport and weather information. It can also provide a conflict probe, which can warn controllers of potential conflicts. Both of these streams of data are manipulated and viewed by the controller using computer equipment known as Enroute Automation Modernization or ERAM.

ERAM

ERAM makes it possible for controllers to access and display information in a customizable format using modern computer equipment. RDP and FDP data are available to all controllers using ERAM. ERAM is a fairly new system that replaces older computer and backup systems that were in use for decades at the ARTCCS. ERAM is the basic platform that will eventually permit the evolution to NextGen, a future ATC concept which will be discussed in more detail in the next chapter. ERAM uses standard commercial hardware and software to provide controllers with a more robust and flexible air traffic control system that can be updated in a more modular fashion.

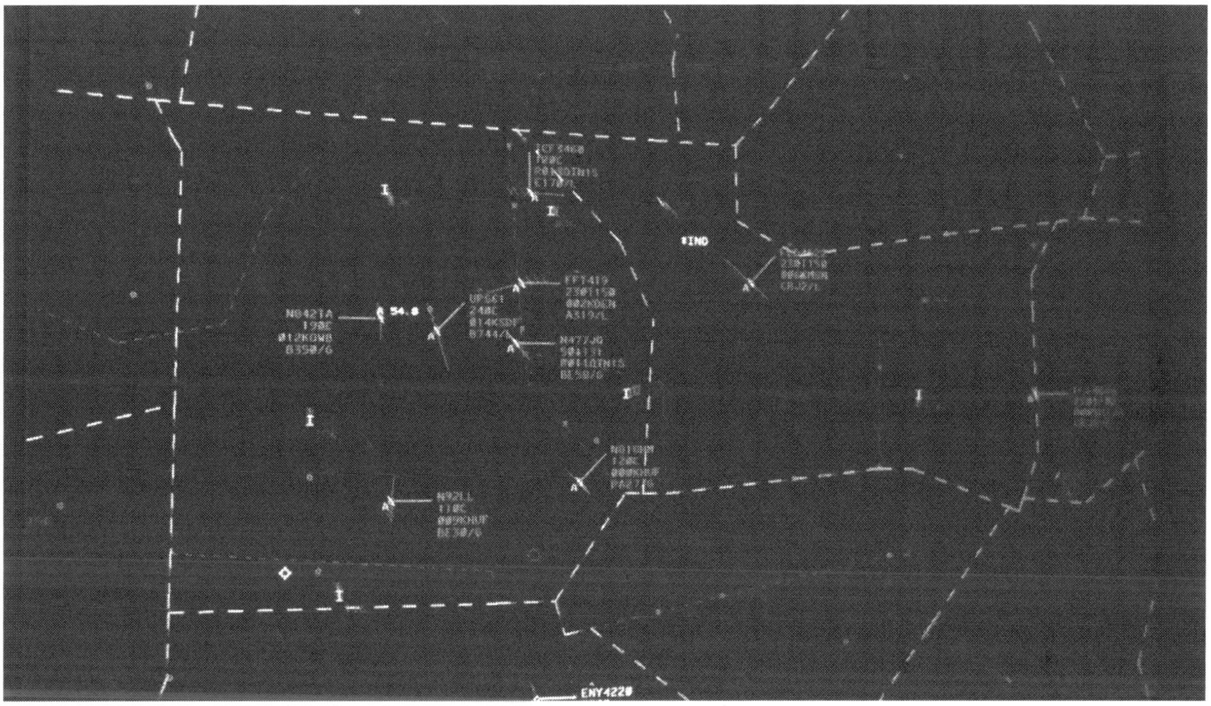

Figure 6-5. ERAM display

ERAM provides the controller with a user-friendly interface with customizable displays. The system is also capable of "trajectory modeling" that predicts the future location of every aircraft, allowing maximum airspace use, and allows controllers to check if proposed changes to an aircraft's altitude or routing will create a conflict, thereby improving the controller's ability to make strategic traffic flow decisions.

Trajectory Modeling

ERAM contains a function that that fuses flight plan and radar data information to create a 4-dimensional trajectory (3D plus time) of every flight from take-off to landing. This capability improves the controller's situational awareness, enables better decision-making providing safer, more efficient routing of aircraft. Trial routings can be tested by the controller prior to issuing them to pilots as well. The system can cross sector and facility boundaries to warn controllers when aircraft might unexpectedly enter their airspace. ERAM will eventually provide controllers with tools that permit dynamic separation and rerouting capabilities, as well as the ability to test more efficient airborne routings around both adverse weather and areas of congestion.

To accomplish these tasks, ERAM combines flight plan and radar track data with aircraft performance characteristics, as well as wind and temperature data to build four-dimensional flight trajectories, for every flight. These trajectories define the horizontal route, altitude profile, airspeed and groundspeed being flown, and the times that the aircraft will reach various points along the route. Each aircraft's actual track is continuously compared to the predicted trajectory position in real time. If the actual track differs from the trajectory prediction in either the lateral, longitudinal, or vertical dimension, a new trajectory is developed.

ERAM uses the trajectories of all aircraft to check continuously for conflicts. A potential conflict will be identified when the trajectories of two aircraft indicate that separation might decrease below acceptable levels. ERAM evaluates the time before conflict, as well as the conflict configuration, to estimate the probability that the current situation will develop into a problem. If the estimated probability is low, and there is adequate time before the conflict might occur, ERAM usually holds off on any notification. Otherwise it notifies the controller of the impending conflict. In general, ERAM looks out about 10-20 minutes into the future--far longer than older legacy systems, which could only project a couple of minutes ahead. In the future ERAM will automatically determine possible solution sets for each situation and will present these possibilities to the controller for selection.

ERAM Display

ARTCC controllers work at an ERAM console equipped with multiple displays that provide both radar and flight plan data. The display consists of various lines, letters, numbers and symbols that aid controllers to separate aircraft. The sector boundaries show the lateral limits of the sectors. Circles denote the location of ground-based navigation stations. Lines extend outward from these navigation stations showing the location of airways. Aircraft are represented on the display as slashes.

Each aircraft has an associated data block that provides the controller with important information such as the aircraft's call sign, altitude, computer identification number (CID), as well as the aircraft's speed across the ground (known as ground speed). If the aircraft is changing altitude, both the current and assigned altitudes are displayed including an arrow indicating whether the aircraft is climbing or descending. The destination airport can also be displayed as well as any other pertinent information that the controller wishes to view.

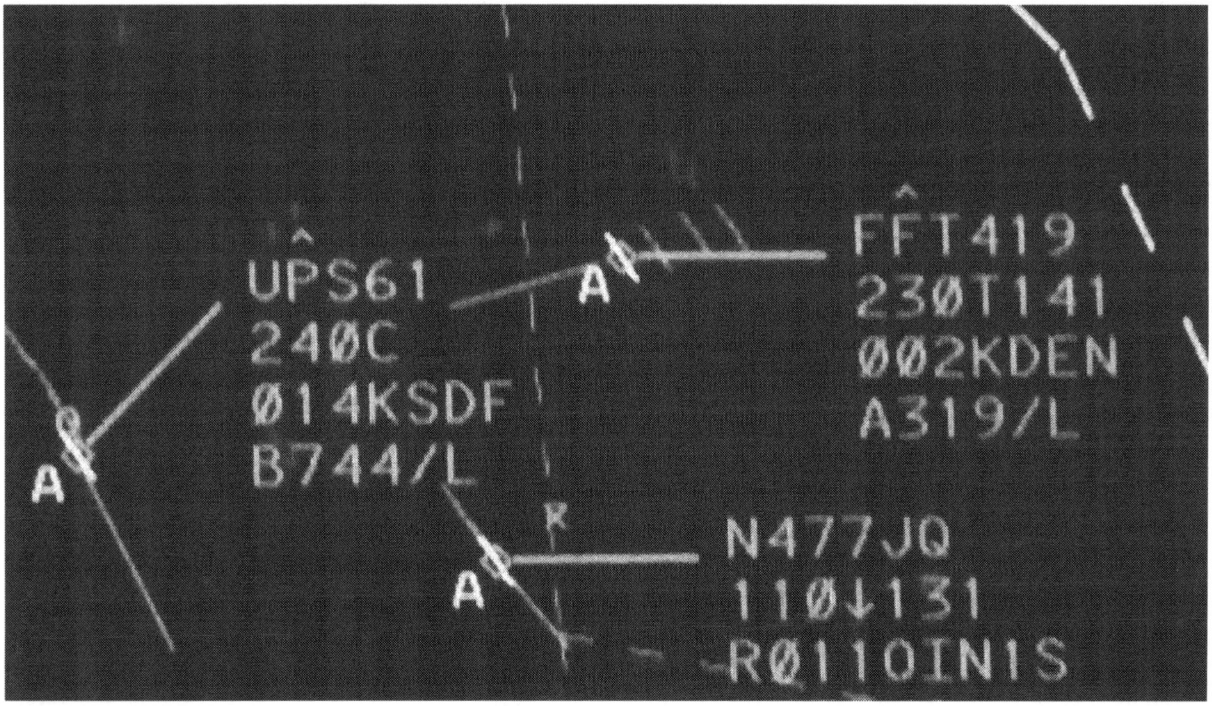

Figure 6-6. ERAM targets and datablocks.

Enroute Separation

Enroute controllers can utilize two different types of separation; radar and nonradar. They usually use radar separation in much the same way as TRACON controllers, issuing vectors, route changes and altitudes to maintain proper separation. Occasionally, they are forced to separate aircraft without being able to see them on the radar; this is known as nonradar separation. Nonradar separation is usually used for aircraft departing and arriving low-activity airports or operating in areas of limited or no radar coverage, such as oceanic airspace. As monitoring systems become more satellite-based, using ADS/B technology, the use of nonradar separation may diminish, but it is still vital for all controllers to understand the basics of nonradar separation as that may be the only available back up system if radars and/or other equipment should fail.

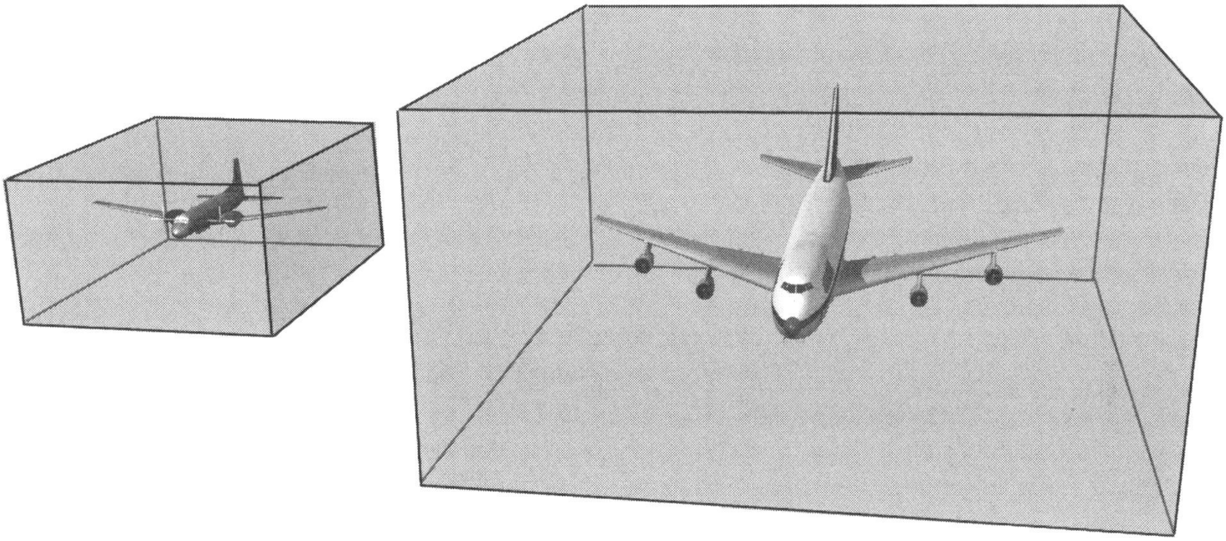

Figure 6-7. Airspace separation

Every aircraft being separated by ATC is assumed to occupy a block of airspace defined in three dimensions; lateral (width), longitudinal (length), and vertical (height). The dimensions vary, depending on whether the controller can "see" the aircraft on radar or whether the position is being determined by the pilot and relayed by radio to the controller (nonradar). Due to the inherent navigation aid inaccuracies previously described, it must be assumed that the aircraft could be located anywhere within their block of airspace and must be protected. The size of each aircrafts "block" of airspace can vary. In general, if positional accuracy is improved, the box can be made smaller. If the controller can "shrink" the block using an appropriate rule, more aircraft can utilize any given area, thereby increasing the throughput and efficiency of that sector. The exact dimensions of the reserved airspace and the procedures applied by the controller vary depending on the aircraft's speed, navigational capability, altitude, and distance from the navigation aid.

Separation Procedures

When separating aircraft, the controller is required to ensure that the airspace reserved for one aircraft does not overlap the airspace reserved for another. If an overlap does occur, even if the two aircraft are in reality miles apart, it is presumed that separation does not exist and that a separation error has occurred. To ensure that the protected airspace assigned to each aircraft does not overlap that of another, specific separation criteria have been developed and must always be used.

Controllers can apply one of three types of separation to aircraft: vertical, lateral, or longitudinal separation. To ensure that the aircraft are in fact separated, the controller needs to apply at least one of these methods at any given time. Vertical separation is based on the altimeter-derived altitude of the aircraft. Lateral and longitudinal separation criteria are based on the aircraft's position, either radar derived, or pilot reported (nonradar). The widespread installation and use of radar has reduced the need for some nonradar applications, but they are still used by controllers whenever radar procedures cannot be applied. In many cases, even radar controllers will sometimes use nonradar criteria for separation since they might be easier to apply in a given situation than radar separation.

The primary principle to be observed when applying separation is that any two aircraft are presumed not to be separated unless separation can be positively proven using one or more of the three methods. The use of any single method is considered proof that separation exists.

Vertical Separation Criteria

Vertical separation is one of the easiest ways to separate two aircraft, whether using radar or nonradar. So long as both aircraft are at altitudes that differ by at least 1,000 feet, they are considered to be separated vertically, as every aircraft's reserved airspace extends from 500 feet above it to 500 feet below. The 1,000-foot standard does vary for aircraft for aircraft operating at very high flight levels. This chart illustrates vertical separation standards:

TABLE 6-4 VERTICAL SEPARATION

Aircraft altitude	Separation required (radar or nonradar)
000-FL410	1000 ft.
FL410-FL600	2000 ft. (special equipment required above FL290
Above FL600	5000 ft.

A controller is allowed to assign an altitude to an aircraft after a preceding aircraft previously at that altitude reports or is observed (radar) vacating it. This rule applies to both climbing and descending aircraft. For example, imagine that N45RM is at 5,000' wishing to climb to 7,000'. A previously departing aircraft, N3WT, had been assigned 7,000', but then received a clearance to climb to 9,000 feet. As soon as N3WT reports (or is observed on radar) leaving 7,000' the controller may climb the lower aircraft, N45RM, to 7,000'. This same scenario would play out similarly if both aircraft were descending.

Lateral Separation Criteria

Lateral separation presumes that both aircraft are on different routes or heading where their reserved airspaces do not overlap. Aircraft separated laterally may operate at the same altitude. Aircraft are considered to be separated laterally whenever at least one of the following conditions exists:

- If using radar, aircraft must be at least 5 nm apart and their courses cannot be converging (getting closer to one another).
  If using nonradar rules;
  - The aircraft must be operating on different airways or routes whose protected airspaces do not overlap. Since each airway is eight miles wide, to be separated laterally aircraft must be operating on airways whose centerlines are at least eight nm apart.
  - The aircraft must be holding over different navigation fixes whose defined airspace does not overlap.

It is fairly easy to determine whether lateral separation exists by using navigation charts. If two airways are at least eight nautical miles apart, they can be used simultaneously by aircraft operating at the same altitude. If the airways are less than eight miles apart, they are not separated laterally. However, even airways that are eight miles or more apart may eventually converge (usually around navaids). The FAA handbook states that lateral separation between aircraft can be considered to exist whenever aircraft are established on different radials of the same navigation aid and at least one aircraft is clear of the airspace reserved for the other. The FAA handbook provides a chart that clarified when lateral airspace overlaps, based on the radials of each airway.

For example, if two airways diverge from each other by 15 degrees, the airspace that belongs to each aircraft overlap until at least one aircraft is 17 nautical miles from the navigational aid. In another example, if the courses utilized differ by 55 degrees, the overlap exists until at least one aircraft is 6 miles away from the navaid. The 7110.65 chart that clarifies these distances and is provided below:

TABLE 6-5 DEGREE DIVERGENCE DISTANCE MINIMA

Divergence (degrees of divergence)	Distance in nautical miles needed for aircraft operating below FL180	Distance in nautical miles needed for aircraft operating at or above FL180
15	17	18
20	13	15
25	11	13
30	9	11
35	8	11
45	7	11
55	6	11
90	5	11

Controllers apply this data by calculating the divergence (in degrees) between the two airways, and then applying the value from the chart to determine where the airways no longer overlap one another. If the value for divergence is not the same as that on the chart, the controller must round up. For example, if two airways diverged from the same VOR by 49 degrees and the aircraft was operating below FL 180, their protected airspaces would be assumed to overlap till at least one aircraft was 6 nm miles from the VOR. The controller would then have to apply some alternative form of separation (usually vertical) while the aircraft were operating within 6 nm of the navaid. As long as one aircraft is at least six or more miles from the navaid, the controller can assume that lateral separation exists.

Longitudinal Separation Criteria

Whenever two aircraft are flying along the same route in the same direction, either vertical or longitudinal separation must be used. Vertical separation is usually easier to apply, but it's overuse will result in an inefficient use of airspace. And at some point, it becomes the enroute controller's duty to line up aircraft inbound for the same airport on the same route, at the same altitude. To accomplish that, the controller must use longitudinal separation.

Longitudinal separation presumes that both aircraft are operating along the same route or are on routes whose protected airspaces overlap one another. For longitudinal separation to be applied, both aircraft must be flying at or near the same airspeed or the leading aircraft must be faster than the following aircraft. Situations in which the following aircraft is faster than the leading aircraft usually make it impossible to apply longitudinal separation. If the following aircraft were indeed faster, it would eventually overtake the leading aircraft, thereby incurring a loss of separation. If the controller must use longitudinal separation, a speed restriction will often be placed on the second aircraft so that it does not overtake the preceding one.

If using radar, the controller must maintain a minimum separation interval of at least 5 nm. It gets much more complicated if the controller cannot use radar separation. Under nonradar rules, in general, if both aircraft are operating at the same speed (or the leading aircraft is faster), the aircraft must be separated by at least 10 minutes of time or 20 nautical miles of distance. If the leading aircraft can be shown to be at least 22 knots faster than the following aircraft, this separation can be reduced to 5 minutes or 10 nautical miles. If the leading aircraft is at least 44 knots faster, the separation interval can be reduced to 3 minutes or 5nm.

So long as longitudinal separation can be shown to exist, aircraft can operate along the same airway at the same altitude. If aircraft are flying on the same route, but headed in opposite directions, vertical separation must be used until the controller can prove that the two aircraft have passed one another. This can be accomplished by having pilots report, (or observing on radar) that both aircraft have passed the same navaid, DME fix, or intersection.

Other forms of nonradar separation procedure do exist for enroute controllers, such as initial departure separation, and arrival-departure separation, but they are not used often. In general, at airports where the Center provides IFR service, controllers usually rely on a "one in, one out" method, where one aircraft at a time will depart or arrive, controllers using vertical separation until both can be are radar identified whereby they can then use radar separation.

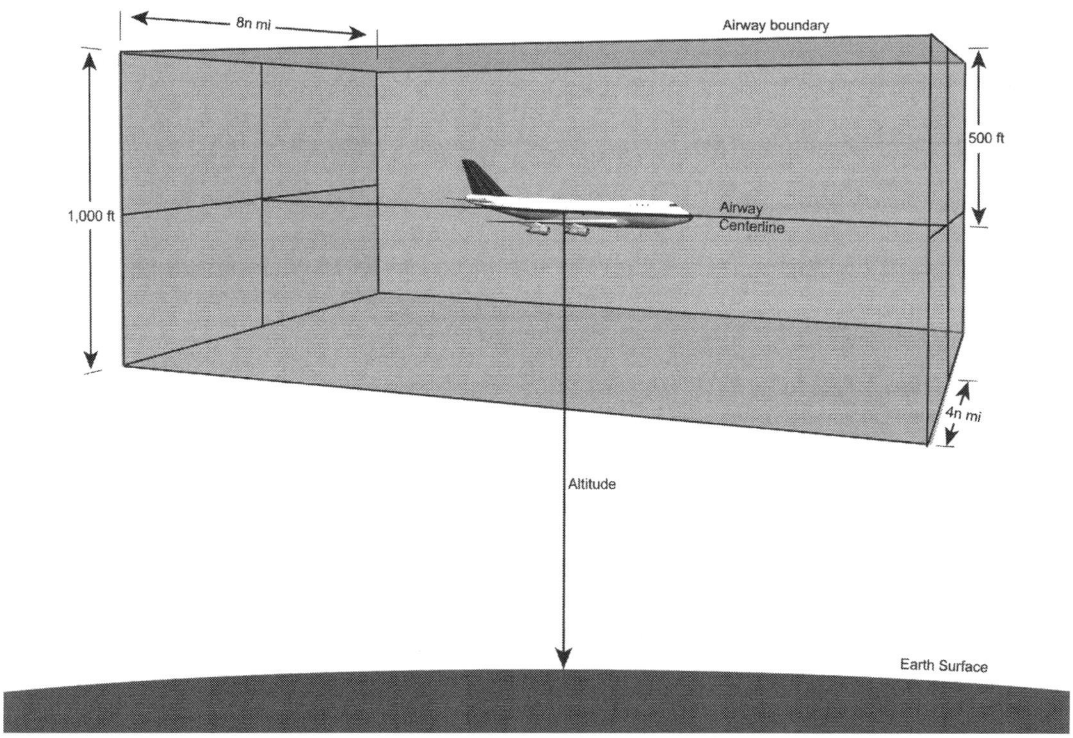

Figure 6-8. Standard non-radar separation minima

TABLE 6-6 ENROUTE SEPARATION STANDARDS

Vertical	Lateral		Longitudinal	
	Radar	Non-Radar	Radar	Non-Radar
1000 ft. Exceptions exist at higher altitudes	5 nm	8 nm (unless the divergence rule is used)	5 nm	10 nm/20 minutes (can be reduced if the leading aircraft is either 22 or 44 knots faster than the following aircraft).

Handoffs, Spacing, and Sequencing

Once an enroute controller accepts a handoff, (either from another ARTCC sector or TRACON), they are responsible for separating that aircraft from all others within their sector. This may be difficult if the aircraft is too low or far away from an ARTCC radar site such that it remains undetected by radar. In these cases, the aircraft will not appear on the ARTCC controllers' radar display and must be separated using nonradar procedures.

Aircraft operating below FL 240 are typically separated by controllers responsible for low-altitude aircraft, known as low-sector controllers. However, if the aircraft climbs to a higher altitude, (above FL 230), it will be handed off to a high-altitude control sector. Once the aircraft reaches its assigned cruising altitude, it continues toward its destination, being handed off to different controllers as it crosses sector and center boundaries. The controllers constantly monitor aircraft separation and make adjustments as needed. If the pilot has any rerouting requests or the controller needs to issue any new routes, ERAM can be used in a trial flight planning mode to determine if any possible conflicts might occur.

As the aircraft flies toward the destination airport, handoffs occur between subsequent sectors. Some will be within the center itself, with others between adjacent ARTCCs. While flying within each sector, it is the controller's responsibility to maintain separation with all other aircraft operating in the sector, avoid military restricted airspace, and comply with all relevant procedures.

Once the aircraft is within 500 to 1,000 miles of the destination airport, traffic flow management programs begin to add to the complexity of the enroute controller's task. If long-term delays are expected at the arrival airport, the aircraft might have been issued a departure delay before the aircraft took off. But, if unexpected weather or other conditions cause a temporary loss of airport or airspace capacity while in flight, aircraft might need to be delayed enroute. There are two basic methods for managing the flow of traffic into an impacted airport--miles in trail restrictions and metering.

FAA traffic management programs attempt to match the inbound flow of traffic to the airport's acceptance rate, which is the calculated rate at which the airport can absorb traffic. If, for instance, calculations show that a particular airport can handle sixty aircraft operations in one hour, its theoretical acceptance rate is one per minute. A general rule of thumb is that a single runway can handle thirty arrivals per hour (one aircraft every two minutes) if the runway is being used for both arrivals and departures. If the runway is being used solely for arrivals, a one-minute interval between aircraft can probably be maintained. However, inclement weather routinely reduces this acceptance rate.

If two aircraft are scheduled to arrive at the runway at the same time, one aircraft will need to be delayed. Such delays place a burden on the approach controller, since only a limited amount of airspace is available to maneuver aircraft within TRACON airspace. It becomes even more difficult to delay aircraft when more than two flights are scheduled to arrive at the same time. In this situation, the approach controller rapidly runs out of airspace within which to maneuver (delay) aircraft.

In general, it is FAA practice to ensure that most of the needed delay be assigned to aircraft while enroute and not while within busy terminal airspace near the destination airport. It is impossible to accurately project all flight paths with minute-by-minute accuracy, so generally it is assumed that, if needed, aircraft can be delayed within TRACON airspace by about five minutes. But, if more than five minutes of delay needs to be assigned, it must be accomplished in ARTCC airspace. FAA procedures require that this delay be imposed far enough out so that when an aircraft crosses an imaginary arc about 200 nm from the destination airport, all of the delay assigned to that flight has already occurred, and the aircraft can begin a continuous descent toward the airport, which is operationally better for both the aircraft and the environment, as it results in fuel savings and emission reductions.

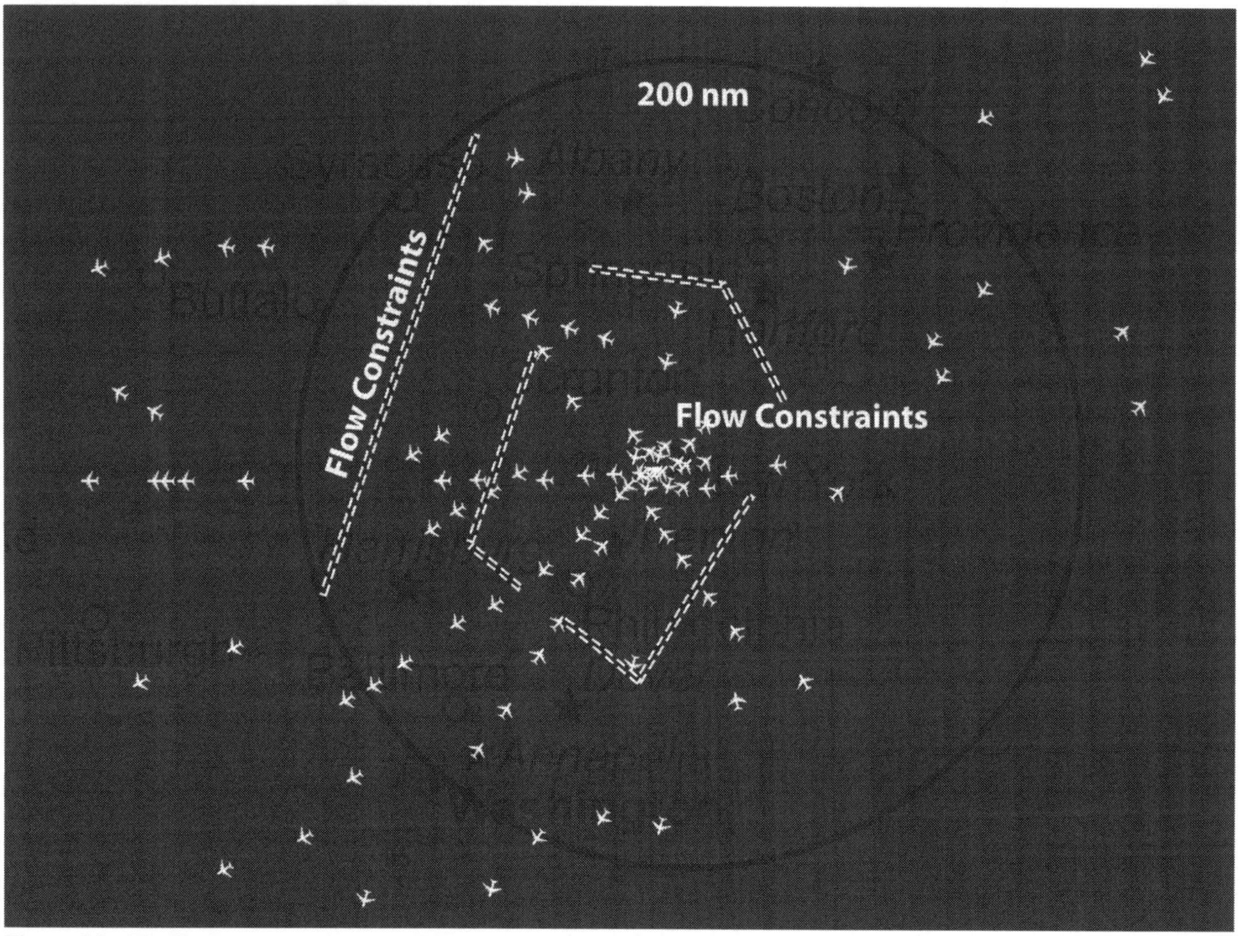

Figure 6-9. Traffic flow constraints and 200 nm arc for the New York area.

Miles in Trail Restrictions

The minimum longitudinal separation for aircraft in enroute airspace is 5 nm. For aircraft flying at about 600 knots, this equates to about 30 seconds of separation. At 250 knots (which is a typical speed for aircraft inbound to the terminal area), 5 nm will be covered by an aircraft in a little over one minute. Assume, for example, that a flow of traffic is flying at 250 knots toward a destination airport from four directions and the airport can safely handle sixty aircraft per hour (or one per minute). If the traffic is essentially evenly distributed and spaced, fifteen aircraft per hour is the limit for the aircraft coming in from each direction. To ensure that the aircraft arrive in an orderly flow, the TRACON would ask the controlling ARTCCs to ensure that the four flows of traffic be evenly spaced with each inbound aircraft spaced about 15 miles apart. This distance is called a miles in trail (MIT) restriction. This request will be communicated between the TMU specialists at the respective facilities, and conveyed to the ARTCC controllers, who would then create the desired spacing.

It is not typical to have equal traffic flow from each direction arriving to an airport however. In real life, traffic flows are not consistent and must be managed in real time. Additionally, more than one arrival runway is typically in use. Common miles-in-trail (MIT) restrictions are usually something like either 10, 15 or 20 MIT. MIT restrictions over corner posts can cause delays throughout the ARTCC since multiple streams of aircraft might be affected. MIT restrictions at an inbound fix tend to ripple back hundreds of miles from the affected airport.

Aircraft Metering

Metering provides similar results as MIT but is a time-based traffic management system. The enroute metering program utilizes the airport's acceptance rate to calculate the number of aircraft that can be handled in any given 5-minute period. If it is determined that the airport acceptance rate will be exceeded, the enroute metering software at the ARTCC begins to issue appropriate delay times to stretch out the line of incoming aircraft.

The metering program dynamically determines times that each aircraft should cross enroute fixes or distance arcs in order to establish the required spacing. It then becomes the ARTCC radar controller's responsibility to ensure that aircraft cross these fixes at the correct times. A rough rule of thumb for enroute spacing is that approximately one minute of delay can be established for every 30 to 50 nautical miles that an aircraft travels. For example, if an aircraft needs to be delayed 10 minutes, the delay needs to start being imposed 300 to 500 miles from the destination airport.

The TMU coordinators at each center calculate the delay to be assigned to every aircraft, then parse that delay out to the sectors. The delay can be displayed as either a time over a fix or a total delay value to be extracted from each aircraft. In the former, a list of aircraft IDs, metering fixes, and times to cross each fix can be placed directly on the center controller's display. It then becomes the controller's job to ensure that the aircraft crosses the fix as close as possible to the assigned crossing time.

Another method involves having the computer display in real time the actual number of minutes that each aircraft needs to be delayed. This number is prominently placed next to the aircraft's data block. Using this system, it becomes the controller's decision how to establish the delay, with the only requirement being that the delay be imposed prior to handing the aircraft off to the next sector or center.

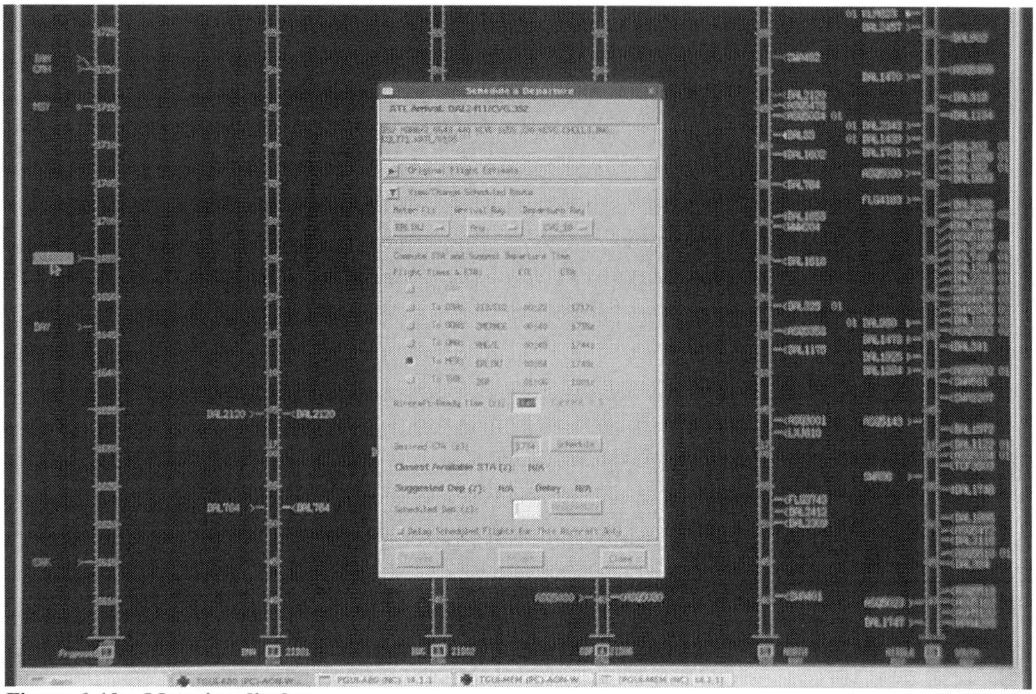

Figure 6-10. Metering display

Delay Techniques

The three most commonly used methods of establishing delay available to the controller include vectoring, speed control, or crossing restrictions. When vectoring an aircraft for a delay, the controller issues a turn that takes the aircraft off course a defined distance and then, at a later time, vectors the aircraft to return to course. This essentially requires the aircraft to fly farther, which will take more time. Another similar method is to issue a new route that is slightly longer than the original planned route of flight.

The controller can also delay an aircraft by slowing it down. This is accomplished by requesting that the pilots either reduce their indicated airspeed by a certain amount, or that the aircraft cross a fix at a certain time. Assigning an aircraft a new speed is somewhat problematic in that the controller sees aircraft groundspeed on the radar display, while the pilot flies in reference to indicated airspeed. At high altitudes and speeds, these two can differ by over 100 knots! Another difficulty with this method is that the controller cannot accurately determine exactly what specific time delay will result from the issued speed change. The controller normally reduces the aircrafts speed until the proper delay distance (or time) has been established. Once correctly spaced, the controller assigns the aircraft an appropriate speed to maintain the proper interval.

The controller could also issue the pilot a crossing restriction, assigning a specific time to cross a particular navigational aid or fix. This requires the pilot to adjust the aircraft's speed accordingly to make the crossing restriction on time. If the aircraft is equipped with a modern flight management system, this calculation and maneuver is fairly simple. Once the aircraft crosses the fix, the controller monitors the aircrafts progress to ensure that the proper spacing interval is maintained.

International Air Traffic Control

Due to their strategic locations, many ARTCC controllers interface with foreign air traffic control facilities as well as control international air traffic. The United States is a member of the International Civil Aviation Organization (ICAO), as are most of the world's countries. Every member of ICAO has agreed to generally abide by common standards unless modified to meet national needs. The adoption of standardized procedures permits pilots to fly around the world using the same language (English), the same navigation aids (VORTAC, ILS, and GNSS), and, at least mostly the same procedures. Without ICAO, every country would be free to develop its own navigation systems, to use its own method for numbering airways and runways, and to use its native language (or languages) for air traffic control. Through the perseverance of ICAO and the cooperation of its members, international air travel is just about as easy as travel within the United States.

ICAO has approved eighteen annexes and three Procedures for Air Navigation Services (PANS) used by member nations when developing and operating their aviation systems. These are as follows:

- Annex 1-Personnel Licensing
- Annex 2-Rules of the Air
- Annex 3-Meteorology
- Annex 4-Aeronautical Charts
- Annex 5-Units of Measurement in Air-Ground Communications
- Annex 6-Operation of Aircraft
- Annex 7-Aircraft Nationality and Registration Marks
- Annex 8-Airworthiness of Aircraft
- Annex 9-Facilitation
- Annex 10-Aeronautical Telecommunications
- Annex 11-Air Traffic Services
- Annex 12-Search and Rescue
- Annex 13-Aircraft Accident Inquiry
- Annex 14-Aerodromes
- Annex 15-Aeronautical Information Services
- Annex 16-Aircraft Noise
- Annex 17-Security
- Annex 18-Safe Transport of Dangerous Goods by Air
- PANOPS-Procedures for Air Navigation Services—Aircraft Operations

Annexes 2, 11 and 14 are those that apply most often to air traffic control. ICAO requires that every country publish manuals describing its ATC system and any differences from ICAO standards. In the United States, this information is contained in the International Flight Information Manual (IFIM) and the Aeronautical Information Publication (AIP). In general, the United States conforms to the recommendations made by ICAO. A few differences should be noted, however.

1. ICAO recommends three types of aircraft operations: VFR, IFR, and controlled VFR (CVFR). Controlled VFR flights are separated by controllers as if they are IFR, but the pilots are not IFR rated and must remain in VFR conditions. Controlled VFR is not used in the United States.
2. ICAO also recommends some phraseology not typically used in the United States. American pilots and controllers pronounce decimal points as "point," whereas ICAO recommends that it be pronounced "decimal."
3. U.S. enroute facilities are known as ARTCCs, whereas ICAO phraseology refers to such facilities as Area Control Centers.

Other than these few minor differences, the U.S. ATC system pretty much conforms to ICAO standards. In recent years, the FAA has made changes to bring its operations more in line with ICAO standards, in an attempt to eliminate confusion among international pilots.

Canadian Air Traffic Control

Because Canada and the United States share one of the longest national borders in the world, the two nations' air traffic control systems interact considerably. This interaction has led to the development of a coordinated system in which both countries have agreed to assist each other in many areas. For example, at some small American airports close to the Canadian border, the FAA has delegated airspace control to the Canadian government. Likewise, parcels of Canadian airspace have been delegated to American ATC facilities.

For this governmental cooperation to come about, both countries' air traffic control systems had to be compatible with one another. Through discussions and agreements between the FAA and NavCanada (the Canadian authority charged with the operation of that nation's air traffic control system), both ATC systems have developed similarly. The procedures used by NavCanada are in many respects similar to those used by U.S. controllers. They are so similar in fact, that in 1985 the United States and Canada signed an agreement recognizing the inherent safety of each other's ATC system. More importantly, the agreement permits the controllers of one country, when authorized to separate aircraft flying over the other, to use the procedures developed by the home country to separate those aircraft.

The area control centers in eastern Canada handle much of the traffic that transits the Atlantic between the U.S. and Europe. A lot of east coast traffic to and from Chicago also transits Canadian airspace. Flights from the midwest to Asia very often spend a lot of time in Canadian airspace as well. A lot of cooperation and data transfer is needed to make these operations seamless, requiring the U.S. and Canadian systems to by similar and compatible.

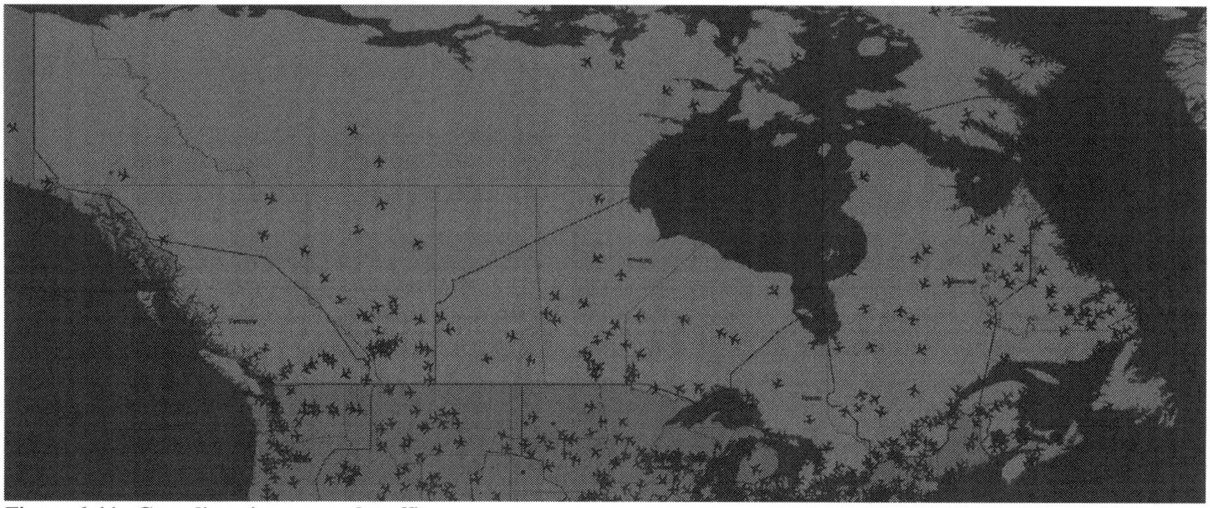

Figure 6-11. Canadian airspace and traffic.

International Airspace

As previously described, ICAO specifies that every nation controls its own sovereign airspace, but international agreements determine which nation shall provide air traffic control service within international airspace, primarily that over the oceans and the poles. Since ICAO is only a voluntary regulatory body and does not provide any direct air traffic control service, international ATC has been delegated to those member nations willing to accept this responsibility. ICAO has assigned a large area of international airspace to the United States. In particular, U.S. air traffic facilities in New York, Oakland, and Anchorage handle large tracts of oceanic airspace.

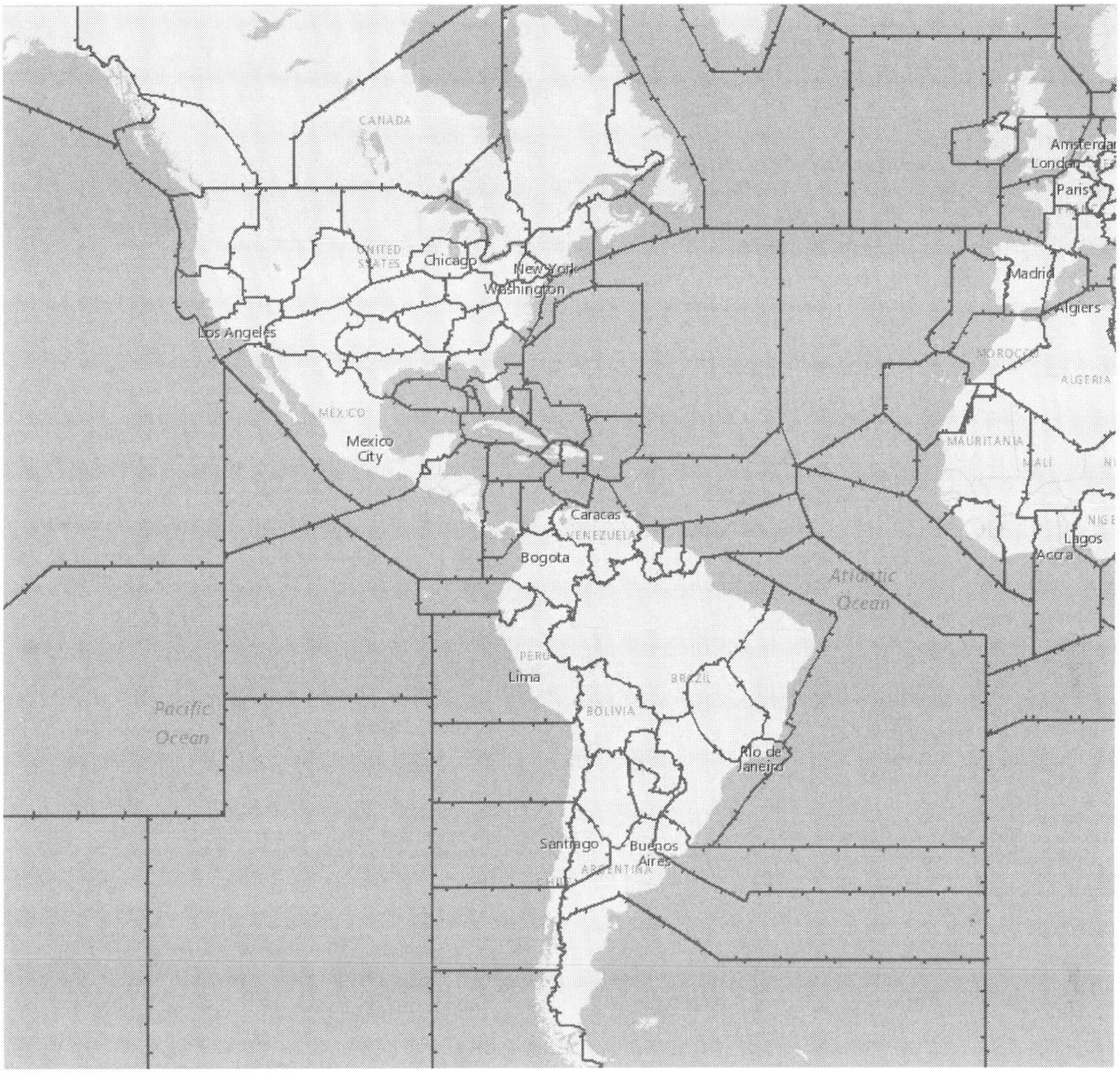

Figure 6-12. Flight information regions of the world.

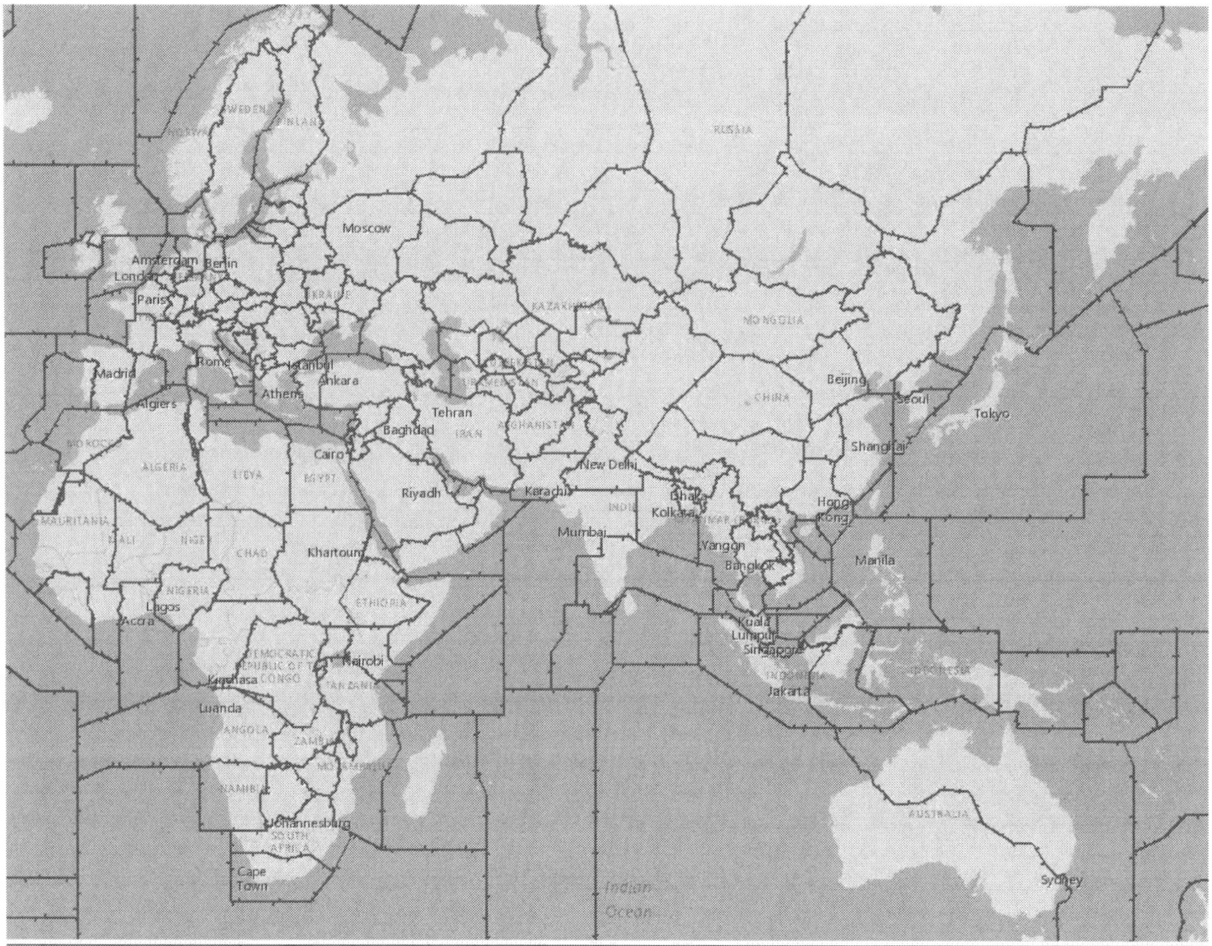

Figure 6-13. Flight information regions of the world.

Although each country with oceanic air traffic control responsibility may not have political ownership of their assigned airspace, the aeronautical regulations of most countries require that their pilots abide by the rules and procedures used by the ACC controlling the oceanic airspace. In the United States, New York, Miami, Oakland, Jacksonville, Houston and Anchorage ARTCCs have significant international airspace under their jurisdiction.

Airport Identifiers

ICAO has established a method of issuing unique identifiers for every commercial airport in the world that differs from that used by the International Air Transport Association (IATA) to identify airports. IATA uses a three-letter format, whereas ICAO uses a four-letter system. IATA's coding format is used primarily by travel agents and airline personnel. The ICAO format is used exclusively in air traffic control.

The ICAO system breaks down the four-letter airport identifier code into three segments. The first segment identifies the area of the world (the aeronautical fixed service routing area or AFSRA) in which the airport is located. The second segment identifies the specific country, and the third segment identifies the airport.

For example, the ICAO airport code of EHAM is assigned to the Amsterdam-Schiphol airport. The letter E in the airport code identifies the airport as being located in northern Europe. The H specifies the Netherlands, and the AM is the code assigned to the Amsterdam-Schiphol airport. In contrast to this system, the IATA code for Schiphol used by airlines and travel agents is AMS.

The only exceptions to this coding scheme are the United States and China. These countries have so many commercial airports that they have unique first letters assigned just to their country: Z for China and K for the United States (although airports in Alaska and Hawaii are identified with the letter P, indicating the Pacific region). The second, third, and fourth letters are assigned to specific airports in these countries. Over 10,000 airport codes are in use worldwide.

Figure 6-14. ICAO airport prefix assignments.

A sample of the airport codes used by the ICAO and IATA follows. A more complete list can be found on many websites including that of ICAO and the FAA.

TABLE 6-7 SAMPLE ICAO AND IATA CODES

ICAO Code	IATA Code	City	Country	Airport
CYMX	YMX	Montreal	Canada	Mirabel International
EDDF	FRA	Frankfurt	Germany	Rhein Main
EGLL	LHR	London	England	Heathrow
KATL	ATL	Atlanta	Georgia	Hartsfield International
PAFA	FAI	Fairbanks	Alaska	International
PHNL	HNL	Honolulu	Hawaii	International
ZBAA	BJS	Beijing	China	Capital

Oceanic Air Traffic Control

The most highly congested international airspace controlled by the FAA is over the North Atlantic Region (NAR). This airspace is congested due both to the amount of traffic and the time zone differences between North America and Europe. Most of the traffic in this region is westbound (Europe to the U.S.) early in the day, and reverses later in the afternoon (U.S. time). Aircraft tend to leave Europe in the morning and arrive in North America in the early afternoon local time. Eastbound traffic is most concentrated between 8:00 P.M. and 3:00 A.M. EST (0100Z and 0800Z), leaving North America in the evening and arriving in Europe early the next morning. Because of this highly directional and concentrated traffic flow, special procedures have been developed for this airspace.

North Atlantic airspace has been delegated to seven different air traffic control facilities. This includes New York Center, with the remainder of the airspace divided among area control centers in Greenland, Newfoundland, Iceland and Great Britain. Until relatively recently, aircraft operating over the north Atlantic were primarily separated using nonradar techniques, since radar service is not available over most of the routes. However, most aircraft transiting the north Atlantic are now equipped with ADS-B (allowing for surveillance), which will be discussed further in the next chapter. Although separation intervals no longer need to be increased, radio communication is still difficult to maintain over this area, which affects the capacity of the North Atlantic airways.

In this region, flights primarily utilize two sets of airways. The first is a series of one-way, relatively low altitude airways, commonly used by single or multi engine propeller-driven aircraft. Aircraft operating on these routes are typically within range of VHF communications facilities and can use VORTAC or NDB for navigation. Most of these airways are designed such that flight over water is reduced to a minimum.

The other set of airways is a flexible system primarily used by airline, military, and business jet operators. The airways are flexible and adjust to varying winds aloft, changing daily based on the wind forecast. These airways can be used only by aircraft equipped with accurate area navigation equipment. Within this airspace, known as minimum navigation performance specifications (MNPS) airspace, only those aircraft properly equipped and certified may operate. Doing so allows reduced separation standards; therefore, the airspace is used more efficiently.

MNPS Airspace Operations

MNPS airspace generally extends from the northeastern United States to Great Britain and from about FL 280 to FL 420. Due to the constraints placed on controllers when separating aircraft within this area, it can become highly congested at peak operating times. To maximize airspace usage, a system of flexible, organized tracks has been developed that replaces the typical airway structure used for air traffic control. These tracks lie between the North Pole and the 27th parallel and between FL 285 and FL 420. The airspace is controlled by New York ARTCC and Shanwick, Gander, Sondestrom, and Santa Maria Oceanic Area Control Centers (OACC).

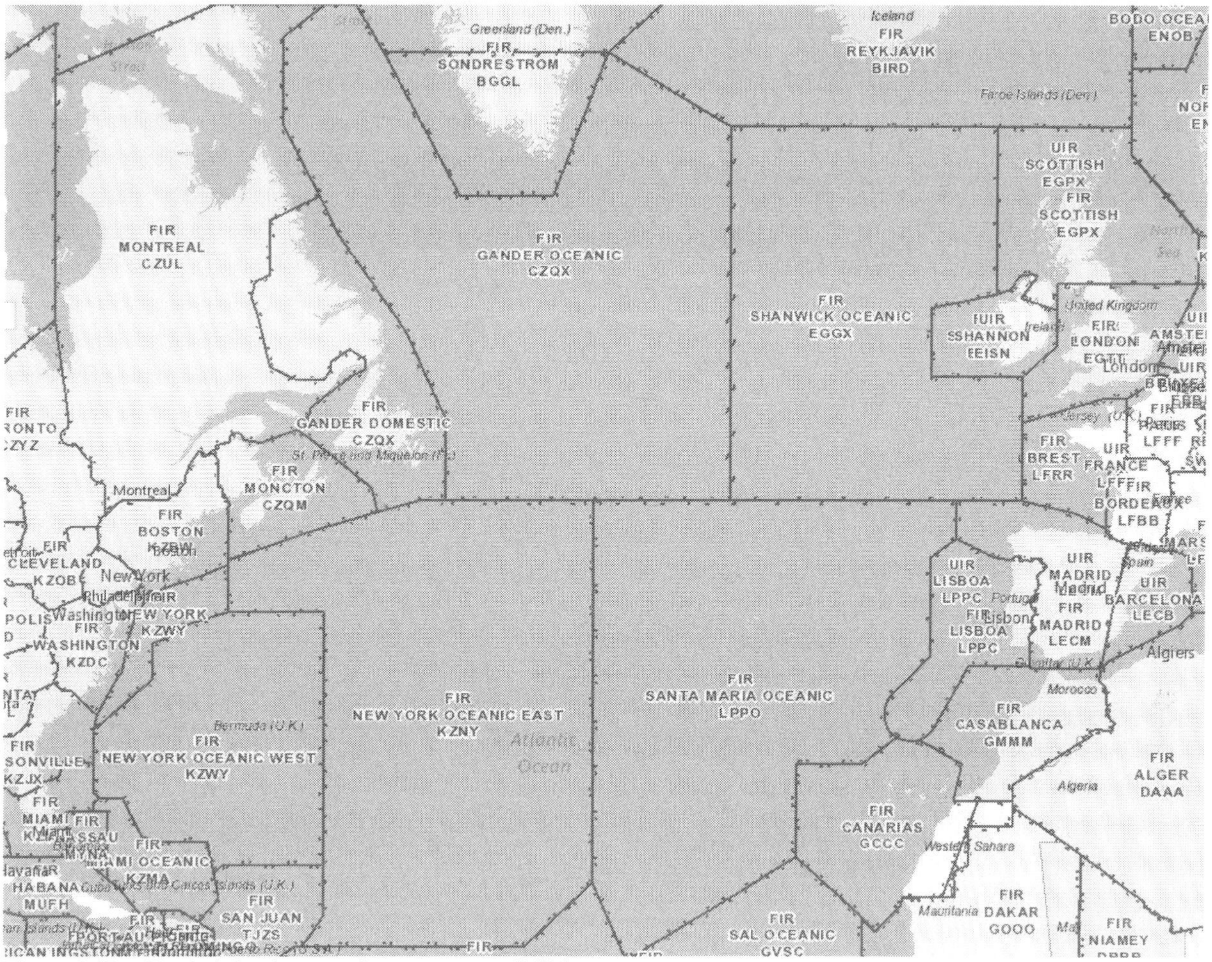

Figure 6-15. North Atlantic control center responsibilities.

To maximize the use of this airspace, international agreements have reduced the separation interval between aircraft operating on these tracks. In return, increased accuracy and reliability of onboard aircraft navigational systems are required. The International Air Transport Association (IATA), in cooperation with ICAO and its member nations, has developed this organized track system and the associated aircraft equipment requirements.

Typically, the organized tracks are developed approximately 24 hours before they are actually to be used. Track development takes into consideration winds aloft and the weather that may be encountered enroute, the anticipated number of aircraft that will be traveling in each direction, and the impact the tracks will have on adjacent and adjoining ATC facilities.

When all factors have been taken into consideration, the organized track system for the next day is provided to potential ATC system users and the ATC facilities themselves. In most cases, two track systems are published. The first is primarily designed for westbound traffic and is effective from 6:00 A.M. to 5:00 P.M. EST (1100Z–2200Z). The second track system is designed primarily for eastbound traffic and is in effect from 7:00 P.M. to 4:00 A.M. EST (0000Z–0900Z). The time interval between these two track systems is used to clear any late aircraft from the system before the tracks are reversed.

Since the actual locations of the tracks change daily, a simple system of naming each track has been developed. This naming scheme informs the pilots of each track location and whether it is eastbound or westbound. The early-morning, westbound tracks are labeled A (Alpha) through K (Kilo). The northernmost track is Alpha and the southernmost is Kilo. The late-afternoon, eastbound tracks are Uniform through Zulu, with the northernmost track being Uniform.

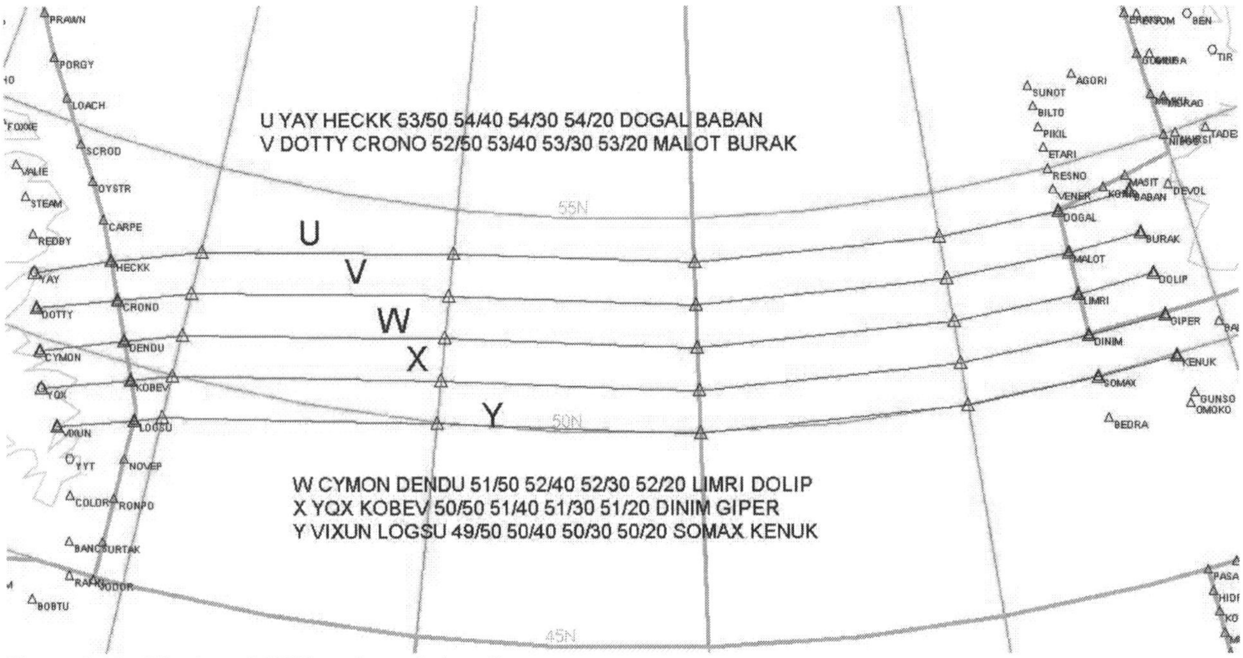

Figure 6-16. Westbound NAT tracks and aircraft.

MNPS Airspace Separation
Aircraft operating along the organized track system can be vertically separated, using the standards and methods previously described in this chapter. However, the primary difference between normal separation and MNPS separation is the application of lateral and longitudinal separation. These reduced separation standards assume that the aircraft are all equipped with ADS as well as controller-pilot datalink communications (CPDLC) equipment.

- **Lateral Separation**-Because of the navigation equipment accuracy required and since aircraft are required to carry redundant navigational systems, lateral separation within MNPS airspace is only 30 nautical miles, which is about half what would be required if the aircraft were operating outside of MNPS airspace.
- **Longitudinal Separation**-Within MNPS airspace, longitudinal separation intervals can be as little as 50 nautical miles or five minutes to properly equipped aircraft which is again about half the separation required if operating outside MNPS airspace. This decreased lateral and longitudinal separation actually permit about four times as many aircraft to use these tracks compared to operating outside MNPS airspace.

Pacific Ocean Air Traffic Control
There are three distinct track areas for traffic transiting the Pacific Ocean. Most of this traffic is controlled either by Anchorage or Oakland ARTCCs. Anchorage controls the traffic transiting from the midwest or northwest of the United States to Asian countries such as Russia, China or Japan. Most of this traffic follows a select series of tracks known as NOPAC. These tracks are parallel spaced and generally lead from the Alaskan region, along the Aleutian Islands, parallel to the east coast of Russia to either China or Japan.

The long distances involved on these routes make wind optimized routing and flight profiles for fuel economy a high priority to users. Westbound routes from New York to Tokyo compete for northern routes which may save significant time by avoiding the jet stream and following a great circle routing. The NOPAC route system comprises five routes that connect Alaska and Japan. The northern routes are generally used for westbound traffic; the southern routes are generally used for eastbound traffic. Similar to aircraft operations over the north Atlantic, aircraft are separated by altitude, route, and speed (mach number) as they leave domestic airspace. Communication with Anchorage ARTCC near the coast is conducted using VHF radio, while over the ocean HF radio is primarily used.

Anchorage ARTCC borders the Tokyo Area Control Center along this route structure. Although this airway system itself does not enter Russian airspace (bordering it just to the south), several recently added tracks now transition Russian airspace, which further decreases distance traveled and thus saves time and fuel.

The Central Pacific (CENPAC) traffic region consists of traffic operating between Hawaii, the west coast of the U.S. and Japan. This region is characterized by long stage lengths and complex weather situations. In this region, the pacific northwest to Hawaii fixed tracks cross the U.S. to Japan routes, creating additional complications for controllers. The traffic flow between Hawaii and the South Pacific (SOPAC) utilizes fixed tracks as well as some random tracks. SOPAC traffic is also characterized by lengthy tracks, such as those from San Francisco and Los Angeles to Australian destinations.

Oakland ARTCC controls the majority of the central and southern Pacific airspace. Oakland is responsible for nearly 20 million square miles of airspace. The Oakland flight information region extends from the west coast of the United States and borders that of Tokyo, Manila, Mexico, Tahiti, Auckland, Nadi, Port Moresby, and Biak. Honolulu ARTCC controls a small section of airspace immediately surrounding the Hawaiian Islands (see Figure 11–5).

Previously, while flying over the Pacific, pilots communicated with the appropriate ARTCC or area control center via HF oceanic radio stations. Aircraft reports, messages, and requests were relayed by the station to the appropriate air traffic control center via telephone, computer, or data message. With the advent of CPDLC however, most aircraft are directly in communication with the controlling air traffic facility via digital messaging systems.

Advanced Technologies and Oceanic Procedures

Within both Atlantic and Pacific airspaces, FAA controllers can monitor aircraft position using advanced technologies and oceanic procedures (ATOPS) equipment. ATOPS integrates flight and radar data processing, accepts manually entered aircraft position information, and displays aircraft location electronically using a graphical display. It also detects conflicts between aircraft, provides a data link, and automates the manual data handling processes previously used by controllers. As ADS-B becomes more prevalent, it will accept aircraft position data from that system as well.

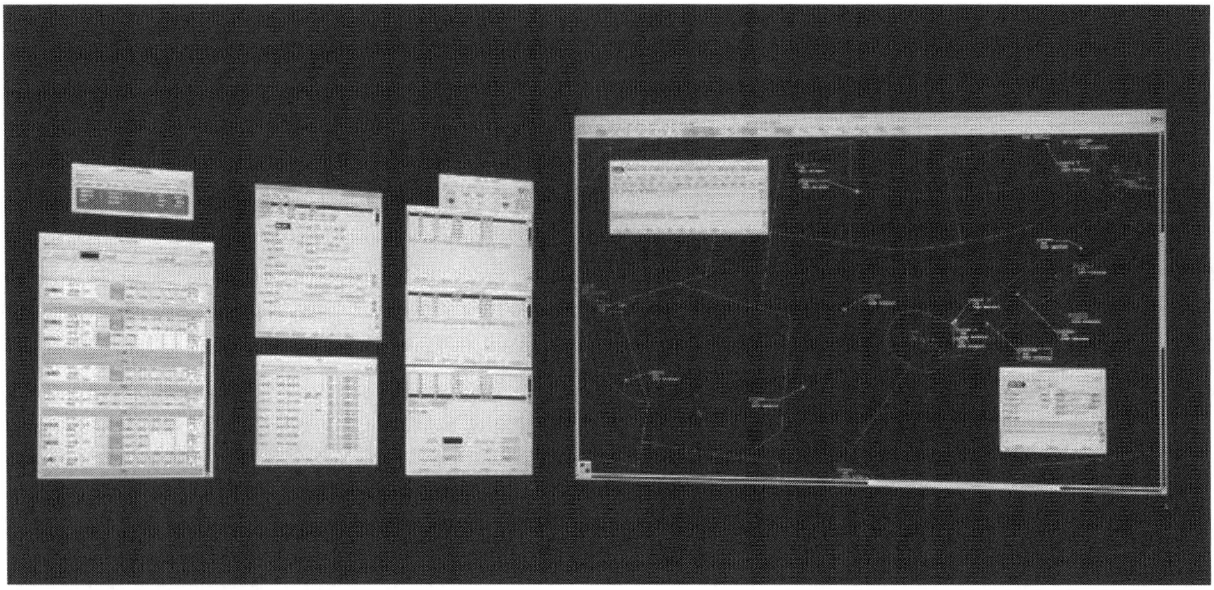

Figure 6-17. ATOPS display

Trans-Polar Flights

As traffic between the United States and Asia increases, the use of trans-polar routes reduces flight times, thereby reducing airspace congestion and increasing airline efficiency. More and more airlines are beginning to use trans-polar routes, but there are some substantial problems associated with their use that must be overcome before they can be fully utilized. To better describe operations at, or near the poles, these flights are sometimes called “high latitude operations”.

Much of the territory covered by high latitude flights offer very few alternate landing areas in the case of an aircraft engine or system problem. In addition, the atmosphere over the poles is particularly cold at cruising altitude, causing fuel flow and freezing problems. Aircraft manufacturers, as well as the airlines, are attempting to reduce these problems through the establishment of a network of emergency landing airfields and navigation aids located in these remote areas.

Fuel temperature problems while in flight are difficult to contend with as it is impractical to heat the large quantities of fuel present in a modern airliner. The fuel placed on board the aircraft prior to departure is relatively warm (ambient air temperature) and might even heat up a little in flight due to air friction across the wings. However, as the fuel quantity decreases and the air temperature at cruise drops significantly, the temperature of the remaining fuel can drop low enough that fuel flow problems develop.

Before departing on a trans-polar flight, flight crews calculate the probable fuel temperature loss that can be expected. The crew monitors the fuel temperature while in flight and may need to make altitude or route adjustments to keep the fuel within the correct temperature range. In extreme cases, flights might need to seek different routes, reverse course, or even land at an alternative field if the flight crew projects that the fuel temperature will become unacceptably low.

Trans-polar operators also need to be equipped with many different modes of communication to ensure that contact can be maintained with ATC during the entire flight. In general, each aircraft must be equipped with very high frequency, high frequency, satellite voice communication as well as data link equipment. The same navigation equipment required for north Atlantic MNPS operation is also required when flying trans-polar.

Trans-polar navigation is quite different from that over the domestic United States. While the departure and initial cruise portions of a polar flight are routine, as the airplane nears the pole, two problems become evident; magnetic compass inaccuracy and meridian convergence. Conventional magnetic compasses sense direction by detecting the horizontal component of the Earth's magnetic field. Since the horizontal magnetic lines of flux become more vertical near the pole, magnetic compasses become increasingly unreliable and unusable in an area approximately 1,000 nautical miles from each magnetic pole. The fact that magnetic north pole and the true north pole are not coincident injects additional uncertainty while navigating. Between these two "north poles," very rapid changes in magnetic variation occur over very small distances.

An additional problem near the poles is the convergence of the meridians or lines of longitude. In conventional navigation, all the lines of longitude converge at both the north and south poles. At either of the poles, there is only one direction to fly, and it takes you wherever you want to go! For example, the only direction available to an aircraft directly over the north pole is south, and that heading would take you just as easily to New York as it would to London or Tokyo.

This convergence of the meridians makes flying traditional great circle routes using magnetic or true headings virtually impossible at latitudes greater than 67° north. At these higher latitudes, due to the convergence of the lines of longitude, very small changes in aircraft position, even with no actual change in heading, can create large changes in both true and magnetic headings.

The solution to this problem is to use inertial navigation system (INS) or GNSS-derived aircraft position and use a modified grid-based form of navigation. Using the prime meridian as a reference, flying "north" from the prime meridian to the North Pole is called "grid north" since it follows the 0 (or 360) degree line of longitude. If a pilot wants to fly from the North Pole to Chicago for example, the direction would be "grid west" as Chicago is approximately 270° west of the prime meridian. This combination of polar grid navigation and true headings and bearings is used until the aircraft is south of about 75° latitude, whereby normal navigation can then resume.

Chapter 7

Future of Air Traffic Control

The U.S. air traffic control system has developed in spurts over the last 50 years. Intervals of relative calm have been repeatedly interrupted by periods of major change and system improvement. Most of these improvements have been made in response to changes in public opinion resulting from major aircraft accidents or incidents. These events have forced the FAA to develop multiple air traffic control system research and development plans.

Almost every time the FAA has modernized, a complete replacement of the ATC system was generally accomplished. Although technically successful, these system upgrades normally took decades to complete and were well over budget. This has not entirely been the FAA's fault; the appropriation and procurement process as well as unforeseeable changes in the aviation industry—such as deregulation, the PATCO strike, and hub-and-spoke airline systems have been partly to blame.

Future Growth in Air Traffic

Although there were declines in air travel in the wake of the 2001 attacks and the recession of 2008, air traffic forecasts indicate a steady growth rate of over ten percent per year for the foreseeable future. As a result, the primary objective of the FAA is to introduce automation and improved reliability into the ATC system that can be upgraded to meet future needs without requiring a complete and total overhaul of the system every time.

In response to Congressional pressure, the FAA began a process around the year 2003 to implement what is now known as the Next Generation Air Transportation System (NextGen). As originally described, NextGen was conceived as a phased implementation of a new national airspace system to be completed by the year 2015 (a date which has passed). As currently envisioned, NextGen is now a continuous process to develop an air traffic control system that by 2025 that will allow pilots more operational flexibility, provide controllers with more management tools, and can be easily upgraded as new technologies become available.

Next Generation Air Traffic Control

In the current air traffic control system, the pilot determines the flight's objectives and how those objectives can best be met. This includes route of flight, proposed altitude, cruising airspeed, time of departure, climb and descent profiles, and speed schedules. A flight plan however, conveys only a limited number of these objectives to the controller; parameters such as speed scheduling and climb and descent profiles are not communicated to ATC.

Currently a controller can query the pilot about these parameters or can make gross determinations of the aircraft's flight profile by interpreting the track and altitude information displayed on the radar scope. Using this limited information, the controller is responsible for separating participating aircraft while also maintaining overall system efficiency. This traffic flow task is further complicated by systemic procedural restrictions that often cause inefficient routing of aircraft, as well as changes due to weather conditions. In addition, at any time during the flight, the pilot may request a change in altitude, route or both which may negate some previous plans and/or traffic flows.

Although the current ATC system attempts to satisfy each pilot's request for a specific route or altitude, procedural restrictions, such as one-way airways or preferred routes, are used to ensure positive aircraft separation while maintaining an efficient, orderly flow of traffic. As the number of aircraft participating in the ATC system increases however, additional route and altitude restrictions must be instituted to reduce potential conflicts between converging streams of traffic. The reality is that imposing these procedural restrictions often separates potential rather than actual air traffic; thus, aircraft may often be denied the use of "empty airspace." Ironically, this may be the very airspace that the pilot had originally requested but was not cleared to use because of an ATC-imposed procedural restriction. The routine use of these restrictions results in increased fuel use and flight times, loss of flexibility and reduced traffic flow.

Since the capacity of today's national airspace system is constrained by numerous rules, procedures, and technologies requiring pilots and air traffic controllers to conduct operations within narrow, often inefficient, guidelines, as air traffic grows, these inefficiencies and their associated costs will be compounded. The current ATC system does not permit pilots to choose either routes or altitudes on any real-time basis. Most traffic management programs "manage" aircraft by delaying or rerouting them. This wastes fuel and increases the time that aircraft are in flight, thereby artificially clogging up the system. Although the current system provides a high level of safety, it was not designed to handle the volume of traffic already present or the predicted annual traffic increase.

In response to these limitations, the FAA and the aviation industry are working together on two major interdependent initiatives—free flight and ATC modernization. According to the FAA, free flight is defined as a concept for safe and efficient flight operating capability under instrument flight rules (IFR) in which the operators have the freedom to select their path and speed in real time. Air traffic restrictions are imposed only to ensure separation, preclude exceeding airport capacity, prevent unauthorized flight through special use airspace, and to ensure the safety of flight. Restrictions are limited in extent and duration to correct only the identified problem; after the conflict has been resolved, aircraft will be allowed to continue on their preferred routing at their requested altitude. Previously, aircraft would have remained on system-mandated routes regardless of traffic or other conflicts; free-flight would remove these types of restrictions.

In theory, a free flight ATC system would allow pilots to choose their own routes, altitudes, and airspeed and modify them in real time as they see fit. Air traffic control would only intervene if necessary to prevent the loss of separation. This system is a far cry from today's ATC system in which pilots are placed on specific routes at predetermined altitudes. Much needs to be accomplished before free flight can become a reality. Additionally, the current system needs to be maintained in some form until the transition to free flight can be accomplished.

Performance Based Operations

Previous air traffic control system components have developed around sets of standards developed by the FAA, ICAO and other aviation organizations. These standards specified every aspect of equipment performance and was necessary for the safe operation of a global ATC system. A major disadvantage of this approach however was the time-consuming process of changing standards to match current technology, followed by the implementation of programs that meet those standards.

Within the last decade, major ATC organizations and providers have slowly shifted to a performance-based system of setting standards and procedures for ATC. Instead of creating a detailed list of standards and polices (that often took an agonizingly long time to implement), these organizations have begun to define outcomes instead of standards. So long as a new procedure, process or piece of equipment can be shown to provide the specified performance (outcome), it can be integrated into the air traffic control system. It is hoped that this new method of definition will permit organizations to become nimbler, and for technology changes to integrate into the ATC system more rapidly and seamlessly. These new standards are being developed in the areas of communications, navigation, surveillance and air traffic management. Hopefully this new approach will work better and speed the advance towards the 2025 goal of full NextGen implementation.

Performance Based Communication

Voice radio is currently the primary electronic method for controller–pilot communications. In oceanic control areas, controllers and pilots already communicate with a version of the controller pilot data link communication (CPDLC) system, which is somewhat akin to texting communications. However, most communications between controllers and pilots currently still occurs verbally over specified frequencies. These radio communications are the essential link for controllers to provide safety of flight instructions. Given the global nature of aviation, aircraft must be able to interface both electronically and verbally with air traffic control systems anywhere in the world.

Today's air traffic control system is overburdened by ever-increasing levels of air traffic and the limitations imposed by congested voice communications, which is known as frequency congestion. This is particularly problematic around busy airports, where the potential for conflicts is greater due to the enormous amount of traffic utilizing a very small amount of airspace. One way to help eliminate frequency congestion is for pilots and controllers to use the correct, succinct approved phraseology.

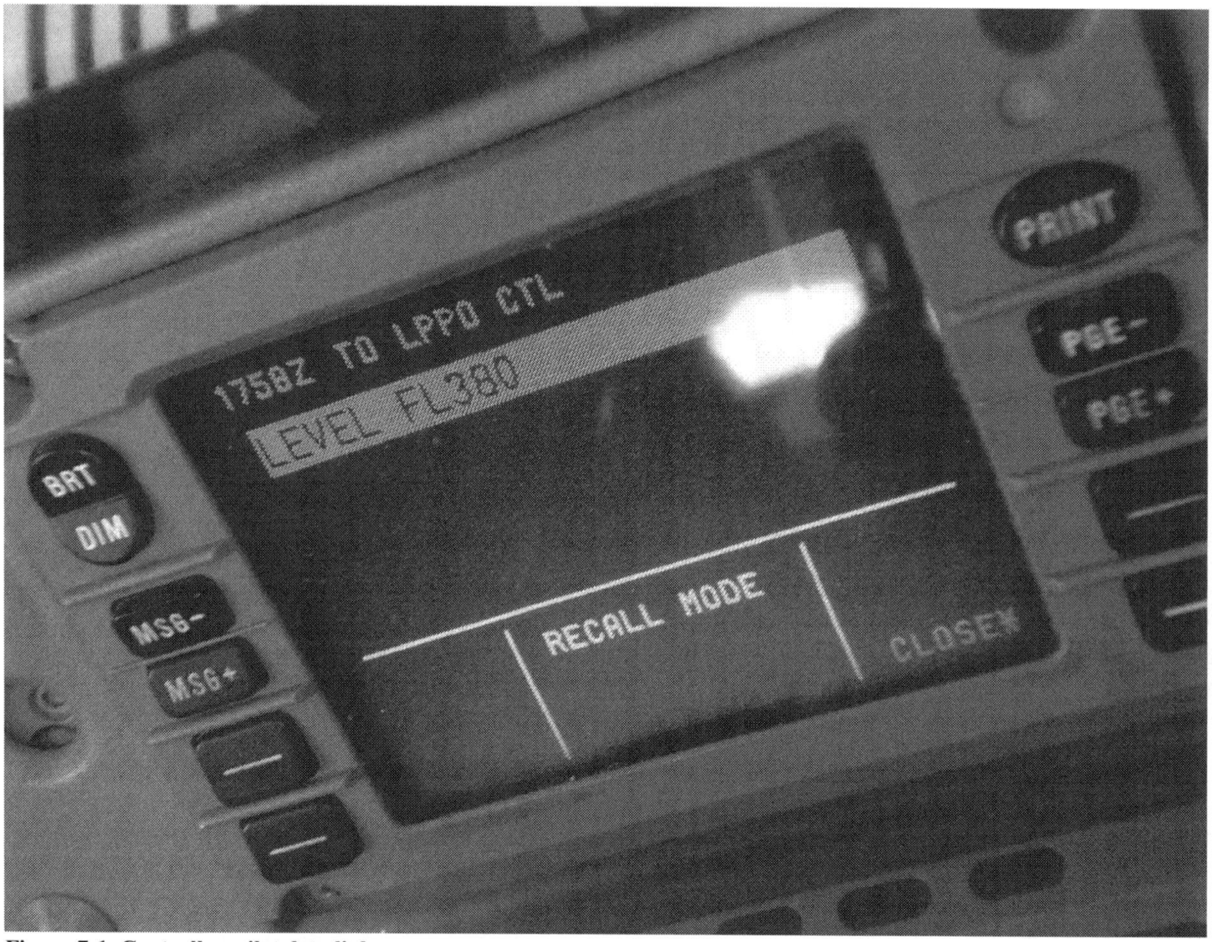

Figure 7-1. Controller–pilot data link.

Performance based communication (PBC) should permit the faster implementation of more modern communications technology. Communication between aircraft and ground facilities in the future will rely less on voice radio communication and more on electronic data transmitted to and from the flight deck via data link technology. Analog radios will be replaced by digital equipment for both voice and data. Digital communication capabilities will be expanded through the use of data link services that should provide a replacement for current air-ground voice communications. After extensive testing, CPDLC was first officially installed in 2002; by 2025, it is hoped that it will replace most verbal communications.

This upgraded system will be accomplished through the implementation of the Next-Generation Air/Ground Communication System (NEXCOM). NEXCOM will include not only voice communication but will also transmit aircraft performance information, air traffic control instructions, weather, traffic and approach information, and other routine pilot-controller requests and communications.

Performance Based Navigation

The FAA is transitioning to a new performance-based navigation (PBN) standard known as required navigation performance (RNP). RNP changes the national airspace system from a ground-based, fixed navaid system to one in which aircraft operators can select which technologies they wish to use for enroute and terminal phases of flight. Aircraft operators will be able to install various independent navigational sensors for their aircraft, choosing from existing VOR, VORTAC, DME, TACAN, GNSS, and/or ILS navigation systems, or even future systems not yet invented.

Once RNP is fully implemented, airspace and procedures will no longer be designed around specific navigational aids and capabilities; instead, RNP technology will define the navigation performance required for aircraft to operate within any particular airspace. RNP, which is based on the minimum acceptable accuracy that can be achieved by the aircraft at least 95 percent of the time, defines the navigation performance accuracy (RNP level) that needs to be achieved for operation within defined airspace or for a particular procedure.

As RNP-certified aircraft fly through the national airspace system, the onboard aircraft navigation system will automatically choose the most appropriate and accurate set of navigation sensors to use and will continuously notify the pilot of the accuracy of the system using RNP levels. This data will be displayed in real time on the appropriate instruments and flight deck displays.

Figure 7-2. RNP navigational display

It is expected that RNP-capable aviation systems will require very little pilot selection of specific navaids or systems. The sensors installed on the aircraft will automatically detect and use the most accurate available system within range and display this information to the pilots. This type of navigation is already commonplace in flight management systems (FMS) found on most modern corporate and commercial aircraft. When using these systems, the aviation suite on board the aircraft detects the most accurate available navigation system or systems, determines the aircraft's location, and displays the aircraft on a moving-map indicator. Little need for pilot input exists, except in the case of an emergency.

Ancillary information, such as which navigation aid is being used and the current accuracy level (RNP level), is also displayed on the screen. When operating on RNP-approved routes or procedures, it becomes the pilot's responsibility to understand his or her aircraft's capabilities and advise the controller if an equipment failure or other malfunction might cause the aircraft to lose its ability to continue operating in the designated RNP airspace.

Performance Based Surveillance

Knowing the position and intended path of aircraft relative to others is necessary to ensure safe separation; as was discussed earlier in the radar chapter. The accuracy and certainty with which aircraft positions can be tracked determine the procedures and spacing used in ATC; thus improved aircraft surveillance can improve the overall efficiency of the nation's airspace by reducing separation requirements. To reduce separation standards while still providing free flight requires the ability to accurately and reliably locate and track aircraft with greater precision and a faster update rate than is used today.

Automatic dependent surveillance (ADS), has been developed to both augment and eventually replace current ground-based surveillance radar systems. Unlike radar, which tracks aircraft using interrogating radio signals, ADS is based on aircraft-derived positioning and transmits reports based on onboard navigational instruments (most likely GNSS). ADS relies on data link technologies to transmit this information.

There are two current forms of ADS; ADS-Broadcast (ADS-B) and ADS-Contract (ADS-C). The ADS-B system broadcasts aircraft location information periodically to all aircraft in the immediate vicinity and air traffic management facilities within a specific area. ADS-C adds the ability to set up an automated "contract" between the ground-based station and specific aircraft. The primary objective of ADS is to improve surveillance coverage, particularly in areas having poor or no radar coverage, such as mountainous or oceanic areas. As ADS usage becomes more widespread, ADS-equipped aircraft within radar coverage will have their positions verified by radar reports, providing independent and redundant surveillance. In areas not covered by radar, ADS will provide the controllers with "radar like" computer displays of aircraft location. Doing so will permit separation requirements in areas with inadequate or nonexistent radar coverage to be reduced. This type of system is already being employed in oceanic airspace, allowing far more efficient airspace usage, resulting in substantial savings for the airlines.

Figure 7-3. ADS and NextGen.

Next Generation Air Traffic Control

In today's air traffic control system, aircraft separation is maintained by air traffic controllers who use radar screens to visualize aircraft flight paths, make subjective judgments as to future aircraft position and potential conflicts, and mentally develop alternate flight paths. Controllers then transmit any required flight path changes via voice radio. Variations in aircraft and navigation performance capabilities are rarely considered, as a single set of equipment-based separation procedures and standards for all aircraft are used. Although many of today's modern aircraft are capable of flying more precise flight paths in both space and time, current flight plans are primarily based on a series of charted airspace points with limited in-flight flexibility in real time.

Some of the major components of the NextGen air traffic management project are as follows:

- Trajectory-based operations
- Flexible Airspace Management
- Collaborative air traffic management

Trajectory-Based Operations

The basis for flight operations under NextGen depends on the development of what is known as trajectory-based operations (TBO). A trajectory can be roughly defined as the four-dimensional flight path of an aircraft through space and time. The concept of TBO is to shift the FAA from operating a clearance-based method of air traffic control to a trajectory-based system of air traffic management. Aircraft operating in the system will be assigned flexible, negotiated trajectories, allowing ATC to morph from a system of individual aircraft control to one of airspace and traffic flow management. The function of controllers will change from that of tactical decision makers to one of strategic traffic flow coordinators. Controllers will manage aircraft trajectories by evaluating overall flows and supervising the adjustment of individual trajectories to ensure aircraft separation. Flexible management of aircraft trajectories will permit maximum utilization of available airspace, providing increased access to the ATC system for all users and allow aircraft with advanced navigational capabilities to fully utilize their capabilities.

While operating in the NextGen environment, aircraft will be required to transmit and receive aircraft and navigational data that includes precise aircraft route information and specific fix crossing times. New surveillance equipment and improved aircraft avionics capabilities will provide controllers and pilots accurate position and trajectory data for each aircraft. Aircraft properly equipped might be authorized to perform some delegated separation that was previously the purview of the air traffic controller. Aircraft operating in the same area with VFR aircraft will have much greater situational awareness. Advanced automation systems will provide the controller with the tools necessary to manage aircraft operating in the system, regardless of their individual capabilities. Automated conflict probes will be available to constantly monitor changing aircraft trajectories, recognizing potential conflicts well in advance, and providing resolution advisories to the controller and pilots.

Controllers will be able to use point-in-space metering to provide safe and orderly traffic flows, increasing the efficient use of airspace. Point-in space-metering will permit controllers to ensure accurate aircraft location over any fix, boundary, or random point in real time. Decision support tools will be developed and implemented that will permit controllers to issue scheduled arrival times over an infinite number of locations. Point-in space-metering will also permit controllers to accurately issue clearances to aircraft in all four dimensions.

Flexible Airspace Management

In the current ATC system, airspace configurations and sector boundaries are mostly inflexible and are based on historical flows and limitations. This inflexibility imposes a capacity constraint on the system during periods of peak demand, airspace restrictions, and weather-related traffic rerouting. While it is currently possible to real-time reconfigure airspace due to changing conditions (such as a runway change at a large airport), the process is unwieldy and rarely used. Generally, the physical size and dimensions of different air traffic control areas, such as facility boundaries or internal sectors, are relatively inflexible. NextGen will enable the ATC system to be flexible enough so that sector boundaries and airspace definitions can be changed in real time to accommodate changes in air traffic flows.

The use of RNP as a navigational standard will enable both pilots and controllers to more efficiently describe and separate aircraft trajectories. The current system of airways and routes very often creates artificial areas of traffic congestion. Busy airports are also constrained by arrival and departure procedures and airspace design based more on navaid and airspace restrictions than the need to separate aircraft or the realities of efficient airspace management. Increased use of RNP will permit the flexibility of point-to-point operations and allow for the development of routes, procedures, and approaches that are more efficient and free from the constraints and inefficiencies of ground-based navaids.

High-density arrival and departure corridors to the nation's busiest airports will be developed to provide a more efficient transition to and from enroute airspace. New arrival procedures that use RNP will be employed at busier airports, allowing for closer route spacing than is available today. Equivalent visual approach procedures will be developed that remove restrictions currently imposed during periods of inclement weather. Precision approaches should become available to every runway, thereby increasing overall airport throughput. RNP-based metering of aircraft both into and out of the terminal environment will provide more efficient use of high-density airspace. Effective use of 4-D metering of aircraft will provide a more efficient use of terminal airspace and airport runways. The overall goal of a newer, more flexible ATC system should be to anticipate and rapidly adapt to changing conditions and environments, rather than slowly reacting to each new technology that come along. As the pace of technology change continues to increase, such institutional advancements are necessary for the nation's air traffic system to remain viable.

Collaborative Air Traffic Management

Collaborative air traffic management (CATM) is an attempt to accommodate aircraft operator preferences to the maximum extent possible, with restrictions imposed only when an actual operational need exists. The concept of CATM is to adjust airspace and air traffic control systems to meet real-time aircraft demand, rather than constraining demand to match system capabilities. If restrictions are still required, the goal of collaborative management is to give the aircraft operator the opportunity to resolve them in a way most advantageous to the operator, rather than the air traffic system arbitrarily defining restrictions. In this way, the air traffic system could move from a static system to a dynamic one, able to adapt to current weather, traffic, and airspace issues.

CATM assumes that all airspace operators desire to work together and collaborate on air traffic management decisions. CATM presupposes a common, inclusive exchange of information to all stakeholders in the air traffic management system, allowing them to collaborate on tactical objectives. Within the CATM environment, aircraft operators will have access to a full range of flight planning tools, including all pertinent ATC data. This will allow them to optimize their flight plans based on internal corporate needs, as well as current system capacity.

The first implementation of CATM, known as Collaborative Decision Making (CDM) is currently operating and was established by the FAA in an effort to improve traffic flow management, establishing closer collaboration between the FAA and system users. All aircraft currently flying under IFR flight plans in U.S. airspace are tracked by ATC computer systems at the ARTCCs. This data is electronically transmitted to the Air Traffic Control System Command Center (ATCSCC) located outside of Washington, D.C. This information is then correlated and verified by the ATCSCC computers. Flights essential to national security or law enforcement are typically removed from the system, with the data then made available to others in the aviation community. This is where flight-tracking websites, like Flightaware derive their information.

However, there are more benefits than just knowing when your relative's flight will arrive; using this information, the FAA and system users can determine actual and projected traffic flows at major airports, monitor flight progress, and make strategic scheduling decisions. See figure 7-4 to see a sample output of some airport arrival data.

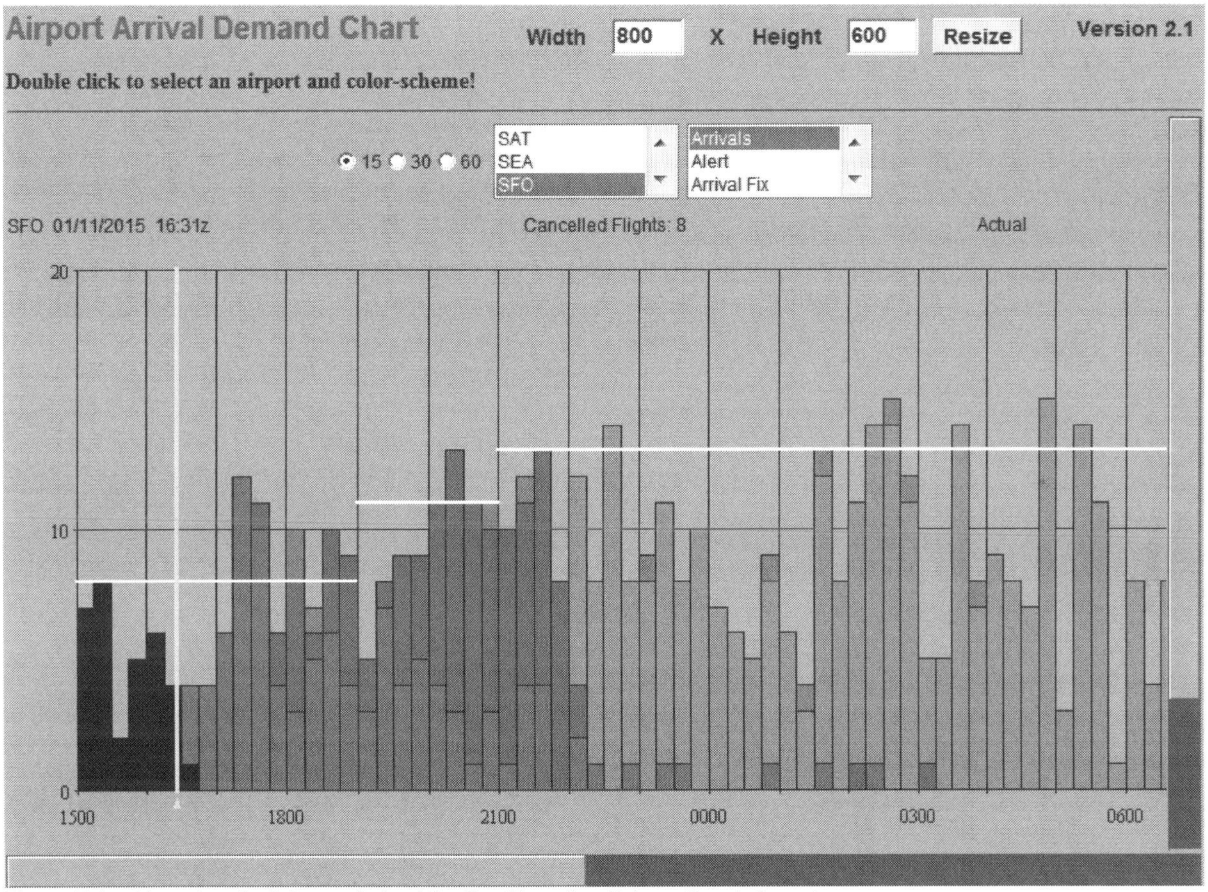

Figure 7-4. Airport arrival demand chart for San Francisco International airport.

The data is currently available to the traffic management units located at every ARTCC and most large TRACONs; many airlines use this data as well. The information is also available to interested systems users through private information vendors or via the internet. These vendors take the raw data provided by the FAA and present the data to users using customized software. Using this information, individuals can visualize traffic flow into and out of airports, predict landings and departures, and even track specific flights. Individual corporations, such as Sabre Airline Solutions, have developed software that permits groups of users to use this data. Large corporations, or even individuals, can gain access to the flight information needed to manage their aviation operations.

As the ATC system becomes increasingly congested, these collaborative programs will become more important to system users. Systems such as these may help eliminate or reduce the scope and duration of ground delays and allow ATC users more flexibility in responding to airport arrival constraints. By using these and other software programs to establish accurate pictures of real-time schedule information, system users will be better able to determine actual and projected traffic flow demands at major airports. They can then balance their company decisions with current ATC system capacity, which should optimize their financial goals. Doing so should correspondingly help the air traffic system by potentially reducing flights at busier times and adding flights during hours that are currently not overburdened.

Future Employment in Air Traffic Control

Every year, the FAA publishes a ten-year strategy for the air traffic control workforce. Much of the data in this chapter comes from the most recent document developed in 2017. This document is available on the internet and is usually updated annually.

The FAA plans to hire somewhere around 1,000 controllers per year for the next five years. The FAA is predicting a slow, gradual decline in hiring during after that that time period however, declining to about 800 controllers per year. Figure 7-5 shows the FAA planned hiring for the next decade. The "peak" shown for 2015 is an aberration and is a "recovery" from the 2013 dip in hiring due to the federal budget sequestration.

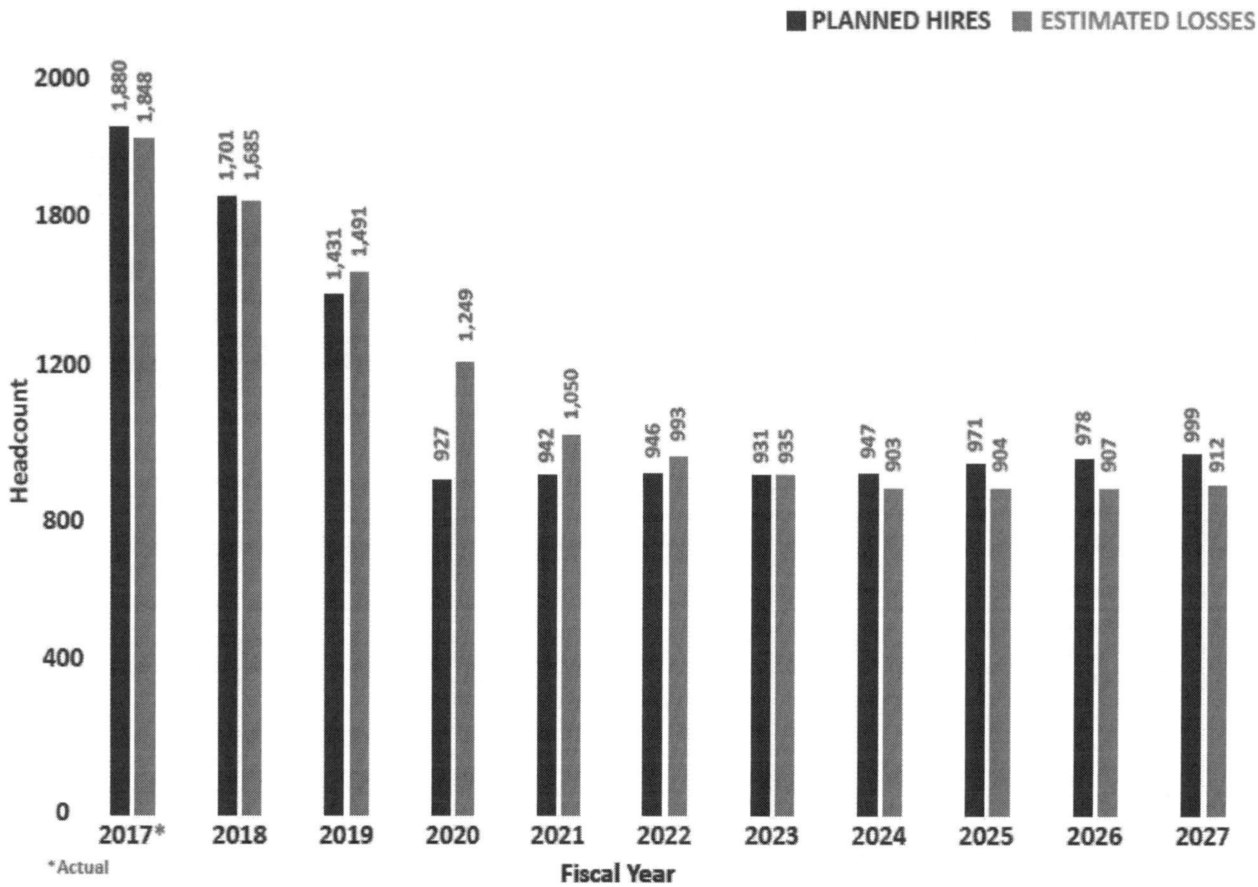

Figure 7-5. FAA hiring projections.

The FAA plans to continue to staff over 200 facilities in the future. In 2016, the FAA had a total of 5,789 enroute controllers on staff; approximately 21% of those are in some stage of training. The FAA also employed 8,258 terminal controllers; again, this number includes approximately 26% developmental controllers. Overall then, the FAA had 14,047 controllers on staff; approximately 24% of these were in some stage of training. According to the FAA, the total number of controllers employed at any time should be between 11,000 and 13,342. The FAA must routinely have more than the allotted number of controllers on staff to compensate for the fact it generally takes from between two and five years for a newly-hired employee to reach full certified professional controller (CPC) status. Although controllers in training can definitely be used to fill staffing needs, doing so reduces their training time, thus delaying their certification as a full CPC.

ATC Privatization

Privatization of the U.S. air traffic control system has been discussed for many years. Self-funded ATC systems are currently operated by Australia, Germany, Canada and the United Kingdom. As far back as 1985, studies were released calling for a self-supporting, non-profit federal ATC corporation in the United States. Although there was little support at the time, the concept was revived by the Clinton administration in 1994 and has been recently brought up in Congress again. In fact, the new administration has said it is one of their main priorities, however, it seems to lack political support in Congress. At this writing, there is no way to predict what will happen.

In past discussions, opposition to any form of privatization was raised by the National Air Traffic Controllers Association (NATCA), as well as organizations representing business and private aviation. Recently, however, NATCA has indicated they would support a privatized system that would allow them to be avoid the budgeting problems inherent in relying on Congressional funding. In recent years, as the FAA has mismanaged many modernization programs, Congress has again begun to study the private operation of an air traffic control system. Also, as political tensions continue to roil the country, it has become increasingly difficult to rely upon Congressional funding to be determined in a timely manner, making it problematic to plan for funding the long-term projects needed to upgrade the ATC system. Privately operated organizations like Nav Canada have shown the capability to operated technically advanced systems while reducing expenses.

The FAA and Congress have experimented with the contracting out of low-activity tower and most of the domestic flight service station system. Although there has not been an increase in accidents or loss of separation incidents at these contracted facilities, debate exists regarding the quality of the services received. Additionally, general aviation (GA) operators are very concerned that in a privatized system, it might cost significantly more for them to operate. They even raise safety concerns, such as whether GA pilots might forgo ATC services due to cost constraints.

Whether the FAA will ever be replaced by a private entity is more a political question than not. It is impossible to predict the future of such an effort, but it could come to pass someday in the future. Individuals considering a job as an air traffic controller need to understand that such an event could change many aspects of the job or have very little effect whatsoever. As with all many things in this fast-paced, exciting job, where it goes remains to be seen.

Figures and Attribution

Figure #	Copyrights and Attribution	Identifier
Title Page	Copyright "James Steidl/Shutterstock"	Image ID: 22884265
Title Page	Copyright "Jirsak/Shutterstock"	Image ID: 46528372
Chapter 1 Title	Copyright "Burben/Shutterstock"	Image ID: 116515978
1-1	Creative Commons, D.B.King	https://www.flickr.com/photos/bootbearwdc/87386298/
1-2	Creative Commons, Federal Aviation Administration	http://commons.wikimedia.org/wiki/File:Potomac_Consolidated_TRACON.jpg
1-3	Federal Aviation Administration	http://tinyurl.com/k8aslvx
1-4	Federal Aviation Administration	http://www.faa.gov/air_traffic/publications/controller_staffing/media/CWP_2014.pdf
1-5	Envato	
1-6	Envato	
Chapter 2 Title	Copyright "Angelo Giampiccolo/Shutterstock"	Image ID: 175280672
2-1	Public Domain - Nolan	
2-2	Public Domain - Nolan	
2-3	Public Domain - Nolan	
2-4	Public Domain - Nolan	
2-5	Copyright "Scott T. O'Donnell/Shutterstock"	Image ID: 670653
2-6	Federal Aviation Administration	
2-7	Federal Aviation Administration	
2-8	eAcademicBooks LLC	
2-9	eAcademicBooks LLC	
2-10	Public Domain	http://www.gps.gov
2-11	Garmin	
2-12	Public Domain	
2-13	Federal Aviation Administration	
2-14	Garmin	
2-15	Federal Aviation Administration	
2-16	Federal Aviation Administration	
2-17	Federal Aviation Administration	
2-18	Federal Aviation Administration	
2-19	Federal Aviation Administration	

Figure #	Copyrights and Attribution	Identifier
2-20	Federal Aviation Administration	
2-21	Federal Aviation Administration	
2-22	Federal Aviation Administration	
2-23	Federal Aviation Administration	
2-24	Michael S. Nolan	
2-25	Federal Aviation Administration	
2-26	Federal Aviation Administration	
2-27	Federal Aviation Administration	
Chapter 3 Title	Copyright "Cylonphoto/Shutterstock"	Image ID: 235841980
3-1	Federal Aviation Administration	
3-2	eAcademicBooks	
3-3	Public Domain - Nolan	
3-4	Public Domain - Wikipedia	
3-5	NASA	
3-6	NASA	
3-7	Public Domain - Nolan	
3-8	Public Domain - Nolan	
3-9	Public Domain - Nolan	
3-10	Public Domain - Nolan	
3-11	Public Domain - Nolan	
3-12	Federal Aviation Administration	
3-13	Public Domain - Nolan	
3-14	Public Domain - Nolan	
3-15	Public Domain - Nolan	
3-16	Public Domain - Nolan	
3-17	Copyright "vectornova/Shutterstock" Copyright "Mechanik/Shutterstock"	Image ID: 240552154 Image ID: 25537183
3-18	Copyright "Keellla/Shutterstock"	Image ID: 205876582
3-19	NATCA	
3-20	Copyright "Copyright: Federico Rostagno/Shutterstock"	Image ID: 46058374
3-21	Copyright "Sander van der Werf/Shutterstock"	Image ID: 222485734
3-22	NASA	
3-23	Garmin	
3-24	Federal Aviation Administration	
3-25	Federal Aviation Administration	
3-26	Raytheon	
3-27	Lockheed Martin	
Chapter 4 Title	Copyright "Andrea Crisante/Shutterstock"	Image ID: 216271171
4-1	Copyright "Artoptimum/Shutterstock" Copyright "Mechanik/Shutterstock"	Image ID: 214021225 Image ID: 25537183
4-2	Copyright "Artoptimum/Shutterstock" Copyright "Mechanik/Shutterstock"	Image ID: 214021225 Image ID: 25537183
4-3	Copyright "Artoptimum/Shutterstock" Copyright "Mechanik/Shutterstock"	Image ID: 214021225 Image ID: 25537183
4-4	Copyright "Artoptimum/Shutterstock" Copyright "Mechanik/Shutterstock"	Image ID: 214021225 Image ID: 25537183

Figure #	Copyrights and Attribution	Identifier
4-5	Federal Aviation Administration	
4-6	Copyright "Artoptimum/Shutterstock"	Image ID: 214021225
4-7	Copyright "Artoptimum/Shutterstock"	Image ID: 214021225
4-8	Copyright "Artoptimum/Shutterstock"	Image ID: 214021225
4-9	Copyright "Artoptimum/Shutterstock"	Image ID: 214021225
	Copyright "Mechanik/Shutterstock"	Image ID: 25537183
4-10	Copyright "Artoptimum/Shutterstock"	Image ID: 214021225
	Copyright "Mechanik/Shutterstock"	Image ID: 25537183
4-11	Copyright "Artoptimum/Shutterstock"	Image ID: 214021225
4-12	Federal Aviation Administration	
4-13	eAcademicBooks LLC	
4-14	Federal Aviation Administration	
4-15	Federal Aviation Administration	
4-16	Copyright "Artoptimum/Shutterstock"	Image ID: 214021225
4-17	Public Domain	http://www.BLS.Gov
4-18	Federal Aviation Administration	
4-19	Federal Aviation Administration	
Chapter 5 Title	Copyright "Herbert Kratky/Shutterstock"	Image ID: 6843121
5-1	Federal Aviation Administration	http://tinyurl.com/ktynyv5
5-2	Michael S. Nolan	
5-3	Image: eAcademicBooks LLC 3D Model:Digimation	
5-4	Copyright "In-Finity/Shutterstock"	Image ID: 216439570
5-5	Michael S. Nolan	
5-6	Michael S. Nolan	
5-7	Public Domain - Nolan	
5-8	Copyright "Artoptimum/Shutterstock"	Image ID: 214021225
5-9	Michael S. Nolan	
5-10	Michael S. Nolan	
5-11	Michael S. Nolan	
5-12	Michael S. Nolan	
Chapter 6 Title	Copyright "Jirsak/Shutterstock"	Image ID: 46528372
6-1	Copyright "boreala/Shutterstock"	Image ID: 271236713
6-2	Federal Aviation Administration	
6-3	Copyright "dikobraziy/Shutterstock"	Image ID: 263958671
	Copyright "Artoptimum/Shutterstock"	Image ID: 214021225
6-4	Copyright "dikobraziy/Shutterstock"	Image ID: 263958671
	Copyright "Artoptimum/Shutterstock"	Image ID: 214021225
6-5	Michael S. Nolan	
6-6	Michael Nolan	
6-7	Image: eAcademicBooks LLC 3D Model:Digimation	
6-8	Image: eAcademicBooks LLC 3D Model:Digimation	
6-9	Copyright "dikobraziy/Shutterstock"	Image ID: 263958671
	Copyright "Artoptimum/Shutterstock"	Image ID: 214021225
6-10	Michael S. Nolan	
6-11	Federal Aviation Administration	
6-12	Michael S. Nolan	

Figure #	Copyrights and Attribution	Identifier
6-13	Michael S. Nolan	
6-14	Creative Commons, Brhaspati	http://tinyurl.com/l7rbln8
6-15	Michael S. Nolan	
6-16	Wikipedia	http://tinyurl.com/k3zcwxp
6-17	Federal Aviation Administration	
Chapter 7 Title	Copyright "Angelo Giampiccolo/Shutterstock"	Image ID: 175363301
7-1	Wikipedia – Creative Commons	
7-2	Garmin	
7-3	Federal Aviation Administration	
7-4	Federal Aviation Administration	
7-5	Federal Aviation Administration	

Index

Aeronautical Charts, 40, 167
air route surveillance radar, 101
Air Route Traffic Control Center, 23, 73, 155
Air Traffic Control Procedures, 71
Aircraft Altitude, 84
Aircraft Flight Plan, 77
Aircraft Identification, 97
Aircraft Metering, 165
Aircraft Speeds, 96
airport surface detection equipment, 101
airport surveillance radar, 101
Airspace, 39, 69, 70, 71, 78, 86, 137, 146, 159, 168, 172, 174, 185, 186
Altitudes, 57, 93
Approach and Landing Procedures, 56
Approach Clearance, 143
Approach Lighting Systems, 64
Arrival Runway Separation, 115
Arriving Aircraft, 112
ARTCC, 85, 86, 126, 150, 151, 152, 153
Canadian Air Traffic Control, 168
Careers in ATC, 14
Category I, 56
Category II, 54, 56
Category III, 55
Class C, 80, 81
Class D, 82
Code Selection, 103
Communications Systems, 90
Conflict Alert, 147
Departing Aircraft Separation, 109
Departure Control, 76, 144
DME, 43, 51, 77
DoD employment, 14
Domestic Enroute Air Traffic Control, 151
Enroute Automation Modernization, 106
Enroute controllers, 15
ERAM, 157, 158
FAA Employment, 15
Final Approach Segment, 50
Flight Levels, 84, 93
Flight Service Stations, 37
Getting Hired by the FAA, 16
Global Positioning System, 45
GNSS, 46, 56, 57, 77
GPS, 45, 46, 51, 55, 56, 58
Ground Control, 74, 122
Hand off Procedures, 73
IFR Separation, 130
Instrument Approach Procedures, 49
Intermediate Approach Segment, 50
International Air TraFfIC Control, 166
Intersecting Runway Separation, 112, 117
LAHSO, 119
Landing Clearance, 114
Local Control, 75, 108
Medical Examination, 19
Minimum Safe Altitude Warning, 148
NRS, 85
Performance Based Operations, 181
Performance Based Surveillance, 184
Precision Approach Path Indicator, 68
Prohibited Areas, 86
Radar, 15, 23, 24, 25, 26, 27, 28, 29, 30, 31, 32, 73, 74, 98, 99, 100, 101, 103, 104, 105, 132, 133, 134, 136, 137, 147, 156
Radar Controllers, 73
Radio Frequencies, 95
Radio Frequency Assignments, 91
Radio Phraseology, 90
Required Navigation Performance, 47
Restricted Areas, 87
Runway Lighting, 63, 64
Salaries, 22
satellite airports, 146
Sectors, 72
Standard Phraseology, 91
Standard Pronunciation, 92
TACAN, 44
Terminal Radar Approach Control, 15, 23
Time, 94
TRACON, 15, 25, 26, 27, 28, 29, 30, 31, 32, 130, 137, 144, 146, 191
VFR, 77, 80, 82, 113, 167
Video Map, 100
Visual Navigation, 39

Visual Separation, 135
VOR, 41, 42, 51, 77
Wake Turbulence, 120, 122
world aeronautical charts, 40